AF692686

Advanced Nanocomposites for Photonics and Optoelectronics and Mechanics

Advanced Nanocomposites for Photonics and Optoelectronics and Mechanics

Guest Editors

Gongxun Bai
Xiewen Wen

MDPI

Basel • Beijing • Wuhan • Barcelona • Belgrade • Novi Sad • Cluj • Manchester

Guest Editors
Gongxun Bai
College of Sciences
China Jiliang University
Hangzhou
China

Xiewen Wen
Department of Industrial
Systems Engineering
The Hong Kong Polytechnic
University
Hong Kong
China

Editorial Office
MDPI AG
Grosspeteranlage 5
4052 Basel, Switzerland

This is a reprint of the Special Issue, published open access by the journal *Nanomaterials* (ISSN 2079-4991), freely accessible at: https://www.mdpi.com/journal/nanomaterials/special_issues/Photonics_Optoelectronics_Nanocomposites.

For citation purposes, cite each article independently as indicated on the article page online and as indicated below:

Lastname, A.A.; Lastname, B.B. Article Title. *Journal Name* **Year**, *Volume Number*, Page Range.

ISBN 978-3-7258-4609-2 (Hbk)
ISBN 978-3-7258-4610-8 (PDF)
https://doi.org/10.3390/books978-3-7258-4610-8

© 2025 by the authors. Articles in this book are Open Access and distributed under the Creative Commons Attribution (CC BY) license. The book as a whole is distributed by MDPI under the terms and conditions of the Creative Commons Attribution-NonCommercial-NoDerivs (CC BY-NC-ND) license (https://creativecommons.org/licenses/by-nc-nd/4.0/).

Contents

About the Editors vii

Preface ix

Hao Li, Esmaeil Heydari, Yinyan Li, Hui Xu, Shiqing Xu, Liang Chen, et al.
Multi-Mode Lanthanide-Doped Ratiometric Luminescent Nanothermometer for Near-Infrared Imaging within Biological Windows
Reprinted from: *Nanomaterials* **2023**, *13*, 219, https://doi.org/10.3390/nano13010219 1

Zoi Georgiopoulou, Apostolis Verykios, Kalliopi Ladomenou, Katerina Maskanaki, Georgios Chatzigiannakis, Konstantina-Kalliopi Armadorou, et al.
Carbon Nanodots as Electron Transport Materials in Organic Light Emitting Diodes and Solar Cells
Reprinted from: *Nanomaterials* **2023**, *13*, 169, https://doi.org/10.3390/nano13010169 12

Ge Song, Zhixiang Liu, Lingchun Jia, Cong Li and Yingli Chang
Modulation of Casimir Force between Graphene-Covered Hyperbolic Materials
Reprinted from: *Nanomaterials* **2022**, *12*, 2168, https://doi.org/10.3390/nano12132168 26

Ting Yu, Rong Luo, Tongbiao Wang, Dejian Zhang, Wenxing Liu, Tianbao Yu, et al.
Enhancement of Casimir Friction between Graphene-Covered Topological Insulator
Reprinted from: *Nanomaterials* **2022**, *12*, 1148, https://doi.org/10.3390/nano12071148 39

Chencheng Zhang, Shengli Pu, Zijian Hao, Boyu Wang, Min Yuan and Yuxiu Zhang
Magnetic Field Sensing Based on Whispering Gallery Mode with Nanostructured Magnetic Fluid-Infiltrated Photonic Crystal Fiber
Reprinted from: *Nanomaterials* **2022**, *12*, 862, https://doi.org/10.3390/nano12050862 52

Cairong Ding, Zesheng Lv, Xueran Zeng and Baijun Zhang
Superfluorescence of Sub-Band States in C-Plane $In_{0.1}Ga_{0.9}N/GaN$ Multiple-QWs
Reprinted from: *Nanomaterials* **2022**, *12*, 327, https://doi.org/10.3390/nano12030327 61

Jialei Miao, Xiaowei Zhang, Ye Tian and Yuda Zhao
Recent Progress in Contact Engineering of Field-Effect Transistor Based on Two-Dimensional Materials
Reprinted from: *Nanomaterials* **2022**, *12*, 3845, https://doi.org/10.3390/nano12213845 72

Chunlan Wang, Yongle Song and Hao Huang
Evolution Application of Two-Dimensional MoS_2-Based Field-Effect Transistors
Reprinted from: *Nanomaterials* **2022**, *12*, 3233, https://doi.org/10.3390/nano12183233 90

Xinxin Ren, Xiang Zhang, Hongxing Xie, Junhu Cai, Chenhui Wang, Enguo Chen, et al.
Perovskite Quantum Dots for Emerging Displays: Recent Progress and Perspectives
Reprinted from: *Nanomaterials* **2022**, *12*, 2243, https://doi.org/10.3390/nano12132243 112

About the Editors

Gongxun Bai

Gongxun Bai, Ph.D., is a Professor at the College of Science and Vice Dean of the College of Science, China Jiliang University. Gongxun received his bachelor's degree from the Huazhong University of Science and Technology, his master's degree from the Shanghai Institute of Optics and Fine Mechanics, Chinese Academy of Sciences, and his PhD degree from the Hong Kong Polytechnic University. Gongxun is the Chief of the National Key R&D Young Scientist Project of the Ministry of Science and Technology and was selected as a distinguished professor of Qianjiang Scholars in Zhejiang Province. Gongxun mainly engages in research on optoelectronic materials and devices and has published over 100 papers in journals such as *Adv. Mater.* and *ACS Nano*, with a total citation count over 8000 and an H-index of 43. Gongxun is currently serving as a member of the Optoelectronic Materials and Devices Special Committee of the Chinese Rare Earth Society.

Xiewen Wen

Xiewen Wen (Sitman) is an Assistant Professor, Presidential Young Scholar, and Associate Director of the University Research Facility in 3D Printing at The Hong Kong Polytechnic University. Prior to joining Poly U ISE, Dr. Wen was an Assistant Professor in the Department of Mechanical Engineering at the City University of Hong Kong and was a postdoctoral fellow at the Argonne National Laboratory in the United States. Dr. Wen was born and raised in Canton and received his bachelor's and master's degrees in physics (optics) from Sun Yat-Sen University before moving to the United States to earn his Ph.D. in Materials Science from Rice University, a top 20 institution in the United States. Wen's group designs and creates one-of-a-kind equipment and materials to create useful manufacturing technologies with ultimate intelligence and precision for applications in optics, electronics, biomedicine, etc.

Preface

In the ever-evolving landscape of materials science, the study of nanostructured materials and nanocomposites has emerged as a cornerstone of innovation. This reprint is dedicated to exploring the fascinating world of these materials, with a specific focus on their photonic, optoelectronic, and mechanical applications.

The subject of this reprint is vast and multifaceted. Nanocomposites, which blend nanosized materials within a standard material matrix, offer unparalleled opportunities for property enhancement. The quantum size effects in nanomaterials lead to significant alterations in their electronic properties, which in turn profoundly modify their photonic, optoelectronic, and mechanical behaviors compared to their bulk forms.

The scope of this collection is broad, encompassing the preparation, characterization, and diverse applications of nanostructured materials and nanocomposites. We aim to delve into the fundamental aspects of these materials, understanding how their structural, mechanical, photonic, and optoelectronic properties interact to give rise to their unique functionalities.

The reasons and motivations behind this reprint are rooted in the rapid advancements and immense potential of these materials in various fields. From nonlinear optics and electro-optic modulators to photovoltaics, plasmonics, lighting, displays, anti-counterfeiting, and photodetectors, nanomaterials and nanocomposites are driving significant technological breakthroughs. However, to fully harness their capabilities, a deeper and more comprehensive understanding of their underlying properties is essential.

This reprint is addressed to a wide-ranging audience, including researchers, scientists, engineers, and students in the fields of materials science, photonics, optoelectronics, and mechanics. It serves as a valuable resource for those seeking to stay abreast of the latest developments in this dynamic area of research.

We are fortunate to have contributions from a group of distinguished authors who have dedicated their expertise to this reprint. Their original research and review articles provide new insights, innovative approaches, and in-depth analyses of the subject matter.

We would like to express our sincere gratitude to all the authors for their hard work, dedication, and willingness to share their knowledge and findings. We also acknowledge the support of the editorial team and the publisher for their invaluable assistance in bringing this reprint to fruition.

We hope that this reprint will inspire further research, stimulate new ideas, and contribute to the advancement of the field of nanostructured materials and nanocomposites.

Gongxun Bai and Xiewen Wen
Guest Editors

nanomaterials

Article

Multi-Mode Lanthanide-Doped Ratiometric Luminescent Nanothermometer for Near-Infrared Imaging within Biological Windows

Hao Li [1], Esmaeil Heydari [2], Yinyan Li [1,*], Hui Xu [1], Shiqing Xu [1], Liang Chen [1,*] and Gongxun Bai [1,*]

[1] Key Laboratory of Rare Earth Optoelectronic Materials and Devices of Zhejiang Province, China Jiliang University, Hangzhou 310018, China
[2] Nanophotonic Sensors & Optofluidics Lab., Faculty of Physics, Kharazmi University, Tehran 15719-14911, Iran
* Correspondence: liyinyan@cjlu.edu.cn (Y.L.); lchen@cjlu.edu.cn (L.C.); baigx@cjlu.edu.cn (G.B.)

Abstract: Owing to its high reliability and accuracy, the ratiometric luminescent thermometer can provide non-contact and fast temperature measurements. In particular, the nanomaterials doped with lanthanide ions can achieve multi-mode luminescence and temperature measurement by modifying the type of doped ions and excitation light source. The better penetration of the near-infrared (NIR) photons can assist bio-imaging and replace thermal vision cameras for photothermal imaging. In this work, we prepared core–shell cubic phase nanomaterials doped with lanthanide ions, with Ba_2LuF_7 doped with $Er^{3+}/Yb^{3+}/Nd^{3+}$ as the core and Ba_2LaF_7 as the coating shell. The nanoparticles were designed according to the passivation layer to reduce the surface energy loss and enhance the emission intensity. Green upconversion luminescence can be observed under both 980 nm and 808 nm excitation. A single and strong emission band can be obtained under 980 nm excitation, while abundant and weak emission bands appear under 808 nm excitation. Meanwhile, multi-mode ratiometric optical thermometers were achieved by selecting different emission peaks in the NIR window under 808 nm excitation for non-contact temperature measurement at different tissue depths. The results suggest that our core–shell NIR nanoparticles can be used to assist bio-imaging and record temperature for biomedicine.

Keywords: fluoride nanocrystals; ratiometric thermometry; lanthanide dopant; upconversion; photothermal therapy

1. Introduction

As a basic parameter, temperature has been widely observed and measured in many fields, such as bioengineering, medical treatment, and physical chemistry [1–5]. Ratio thermometers are based on the multiple emissions provided by phosphors, usually with significant relative sensitivity, S_r. The dual-emission ratio luminescence thermometer shows the resolvable emissions of two different emitters and can be used as a self-reference ratio [6,7]. Temperature-sensitive signal changes can be measured using ratio intensities rather than absolute photoluminescence intensities to reduce the influence of external factors [8]. On the one hand, the high accuracy and reliability of fluorescent nanoparticles in the application of ratio optical temperature measurement can be achieved [9]. On the other hand, lanthanide-doped nanoparticles can generate upconversion luminescence under the excitation of an external excitation light source [10]. Therefore, noninvasive, non-contact biological tissue temperature measurements based on lanthanide-doped nanoparticles are possible [11].

In fact, in the field of non-contact thermometry, lanthanide-doped nanoscale temperature probes can achieve NIR emission at the same time through NIR excitation [12]. Compared with luminescence thermometers based on other materials, lanthanide-doped nanoscale temperature probes can have richer NIR emission bands by changing the doping elements, which provides more options for multi-mode temperature measurement [13–15]. Different from visible light, the absorption and scattering of NIR photons in in vivo environments are significantly reduced. This avoids possible tissue damage caused by UV light, providing a greater reading of tissue penetration depth and optimizing the signal-to-noise ratio [16,17]. The excellent penetrability of the NIR region to biological tissues determines its potential in biological applications, especially for the second biological window (NIR region 2). The second biological window is not only convenient for multi-mode optical measurement but also coincides with the NIR imaging window. This gives the second biological window full use of the NIR-II window (1000–1350 nm) for in vivo fluorescence imaging. This advantage is not available in conventional visible light fluorescence imaging and the NIR-I window (700–950 nm). The imaging quality is greatly improved due to low self-fluorescence and reduced scattering [18–20].

In this paper, the core–shell lanthanide-doped nanoparticles are designed as multi-mode NIR nanothermometers, and their applicability in temperature sensing applications is demonstrated. Compared with 980 nm, 808 nm excitation is more difficult to be scattered and absorbed by water in tumor tissues, and NIR emission involves the NIR-II window and NIR-I window [21–23]. Therefore, by using the two windows, not only multi-mode optical temperature measurement of tissues at different depths can be realized, but they also can be applied for deep biological NIR tissue imaging. Nanoparticles possess two optical properties, temperature-dependent thermal coupling energy level and phonon-assisted thermal sensitivity fluctuation [24–26]. These optical properties make $Nd^{3+}/Yb^{3+}/Er^{3+}$-doped Ba_2LuF_7 nanocrystals have the potential to work in biological tissues as ratio optical thermometers. Their maximum thermal sensitivity is 0.63% K^{-1}, which is superior to some lanthanide-doped fluoride nanoparticles and quantum dot materials, as shown in Table S1 [27–29]. The nanocrystals have certain absorptivity and reflectance for NIR excitation sources, as shown in Figure S1. At the same time, multiple thermal coupled energy levels can be obtained in a single probe by observing emissions in different regions and selecting different emission ratios, to realize multi-mode temperature measurement [30,31]. The novelty of this work lies in the realization of multi-mode optical temperature measurement using different NIR windows and taking into account the imaging of biological windows, which provides new support for the application of the lanthanide-doping nano platform in tumor detection and treatment.

2. Materials and Methods

2.1. Sample Fabrication

Nanocrystals with the nominal chemical formula of $Ba_2Lu_{1\text{-}x\text{-}y}F_7$: zEr^{3+}/yYb^{3+} xNd^{3+} were prepared from trifluoroacetate materials. The values of x, y, and z are 0.03, 0.24, and 0.03, respectively. The precursor of lanthanide trifluoroacetic acid was prepared from the corresponding lanthanide and $BaCO_3$ and trifluoroacetic acid (99%). The raw materials were composed of Lu_2O_3, $BaCO_3$, Er_2O_3, Nd_2O_3, Yb_2O_3, and La_2O_3. The purity of the above oxides is 99.99%. The nanomaterials were fabricated by thermal decomposition. In the case of the Ba_2LuF_7: 3%Nd^{3+}, 24%Yb^{3+}, 3%Er^{3+} triple-doped samples, 0.015 mmol of Er_2O_3, 0.35 mmol of Lu_2O_3, 0.12 mmol of Yb_2O_3, and 0.015 mmol of Nd_2O_3 were dissolved to transparency in 10 mL trifluoroacetic acid solution, where the concentration of the trifluoroacetic acid solution was 50% at 105 °C. The temperature rose to 120 °C and slowly evaporated the remaining water and acid until it left a dry solid. After that, the trifluoroacetate was mixed with 11 mL octadecene and 7 mL oleic acid in a three-necked flask. The purity of oleic acid and octadecene here is 90%. The solution was heated to 130 °C in a nitrogen-filled flask and stirred quickly for 90 min to remove the remaining oxygen and water. It was then heated to 300 °C under dry nitrogen at 10 °C/min and kept at this temperature for 2 h. All the above chemical materials were obtained from Aladdin in Shanghai.

2.2. Characterization of Sample

The X-ray diffraction (XRD) data were collected by X-ray diffractometer (D2 PHASER, Bruker, Germany) with Cu-Kα (1.5406 Å) radiation. The elemental composition of nanocrystals was investigated by X-ray photoelectron spectroscopic analysis (XPS, K-Alpha, Thermo Scientific, WWLP, USA). The surface morphology and particle size distribution of nanocrystals were recorded by a transmission electron microscope (TEM, Tecnai G2 F20, FEI, OR, USA). The optical temperature measurement platform built by us not only recorded the temperature detection performance of the prepared nanocrystals but also evaluated their photothermal performance. The fluorescence spectrophotometer type QM8075-11(HORIBA, Japan) was used to record the photoluminescence spectra of nanomaterials. The ultraviolet absorption spectrum and Fourier transform infrared spectrum of nanoparticles were obtained by an ultraviolet spectrophotometer (UV-3600, SHIMADZU, Japan) and a Fourier transform infrared spectrometer (iS50, Nicolet, WI, USA), respectively. The NIR imaging of nanoparticles was performed by a NIR camera in a dark chamber (3200 OV1080P, Angeleyes, ShengZhen, China). In the absence of additional instructions, the above experiments were carried out at room temperature.

3. Results and Discussion

3.1. Phase Structure and Morphology of Ba_2LuF_7@Ba_2LaF_7 Nanoparticles

The morphology and size of the compound were measured by transmission electron microscope (TEM). As demonstrated in Figure 1a,b, the Ba_2LuF_7@Ba_2LaF_7 nanocrystals were composed of a large number of cubic phase nanoparticles, about 10 nanometers in length. Figure 1a–c show the morphological characterization of nanocrystals at different scales. Based on 100 nanoparticles, Figure 1d also reveals that synthesized Ba_2LuF_7@Ba_2LaF_7 nanoparticles were uniformly sized with mean particle sizes of 10 ± 2 nm. Moreover, the structure of the synthesized nanocrystals was confirmed by X-ray diffraction (XRD) in Figure 1e. According to the half-width of the three strongest diffraction peaks of the XRD pattern, the average diameter of the nanoparticles was about 12.7 nm, which was calculated by the Debye–Scherrer formula. The calculated results were found to be close to the TEM analysis. The XRD patterns for the synthesized Ba_2LuF_7@Ba_2LaF_7 nanoparticles (Figure 1a) confirmed that the samples are crystalline with the diffraction peaks indexed to Ba_2LaF_7 (PDF # 49-0099). The diffraction peak of the Ba_2LuF_7@Ba_2LaF_7 nanoparticles was slightly offset from the Ba_2LaF_7 (PDF # 49-0099), which was due to the substitution of La^{3+} ions by Lu^{3+} ions in the lattice position, forming the cubic phase Ba_2LuF_7. For the sake of deeply comprehending the impact of the $Yb^{3+}/Nd^{3+}/Er^{3+}$ ions doping on the crystal structure of the studied samples, the Rietveld refinements of the typical Ba_2LuF_7: $Yb^{3+}/Nd^{3+}/Er^{3+}$@Ba_2LaF_7 nanoparticles based on their XRD data were carried, as depicted in Figure 1f. As expected, these calculated diffraction bands were identical to those of the experimental data, which implied that the resultant nanoparticles exhibited a pure cubic phase [32]. As shown on the upper right in Figure 1c, the high-resolution TEM image consisted of clear lattice fringes with a spacing of around 0.213 nm.

3.2. XPS Analysis of Ba_2LuF_7 and Ba_2LuF_7@Ba_2LaF_7 Nanoparticles

The chemical composition and electronics state information of Ba_2LuF_7 and Ba_2LuF_7@-Ba_2LaF_7 nanoparticles were investigated by X-ray photoelectron spectroscopic (XPS) analysis. Figure S2a,b describe the recorded XPS spectra of Ba_2LuF_7 and Ba_2LuF_7@Ba_2LaF_7 nanoparticles, respectively. In addition, these XPS survey spectrum peaks verified the presence of Ba, Lu, La, and F aspects. Additionally, the C1s peak located at a binding energy of 285 eV might be due to the XPS instrument's adventitious hydrocarbon. In Figure S2c, The Ba 3d deconvoluted spectrum shows the binding energy peak, and 780 eV for Ba 3d spin-orbit components is fairly similar to the theoretical value of Ba^{2+}. Figure S2d–f core level spectra show the peaks on La 3d, Lu 4d, and F 1s at 836, 198, and 684 eV, respectively [33]. This proved the existence of Ba_2LuF_7 and Ba_2LaF_7. Therefore, the XPS studies confirmed that Ba_2LuF_7@Ba_2LaF_7 nanoparticles were obtained without foreign contaminants.

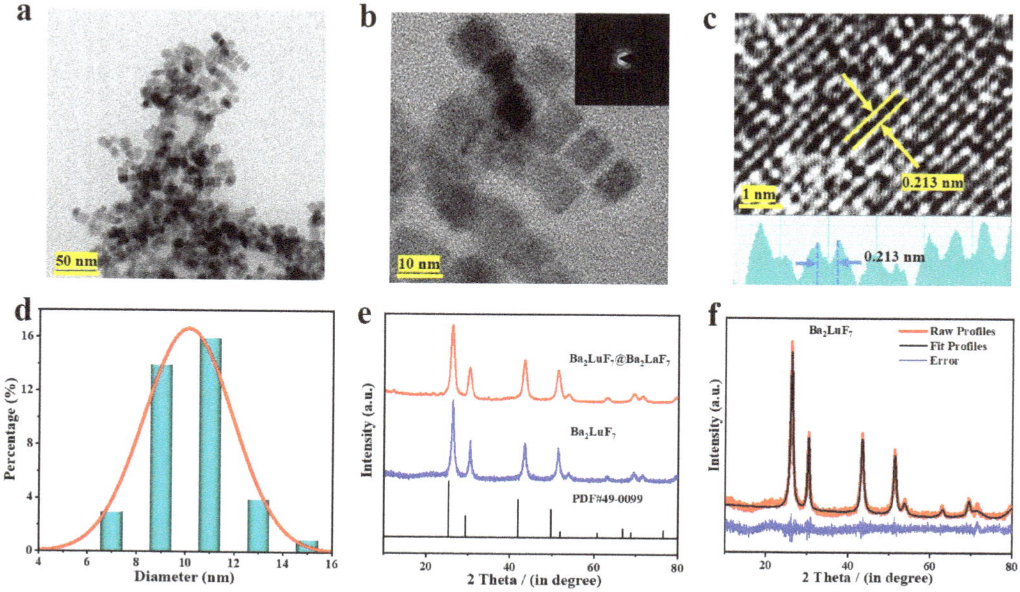

Figure 1. (**a**) Morphological characterization of nanocrystals at 50 nm scales was obtained by TEM. (**b**) Morphological characterization of nanocrystals at 10 nm scales was obtained by TEM. The selective area electron diffraction (SAED) pattern is presented in the illustration on the upper right. (**c**) The high-resolution TEM (HR-TEM) image of Ba_2LuF_7@Ba_2LaF_7. (**d**) The particle size distribution plot for Ba_2LuF_7@Ba_2LaF_7. (**e**) XRDs of tri-doped Ba_2LaF_7 and Ba_2LuF_7@Ba_2LaF_7 nanocrystals were compared with those of standard cards. (**f**) Rietveld XRD refinement for Ba_2LuF_7@Ba_2LaF_7 nanocrystals.

3.3. Photoluminescence Properties of Nanoparticles

We investigated the near-infrared upconversion photoluminescence properties of the Ba_2LuF_7: Nd^{3+} /Er^{3+}/ Yb^{3+} system based on the spectral data of the nanoparticle system. The photoemission spectrum of the prepared tri-doped Ba_2LuF_7 nanoparticles under 808 nm photoexcitation is shown in Figure 2a. The curve shown in Figure 2a shows that the upconversion emission performance of all samples is consistent with the reported performance. It consists of four strong peaks triggered by the Yb^{3+}/Er^{3+}/Nd^{3+} transition, and the central wavelengths are 974 nm, 1052 nm, 1321 nm, and 1527 nm, respectively. These transitions are reported to correspond to the $^2F_{5/2}\rightarrow^2F_{7/2}$ transition of the Yb^{3+} ion. This proves that Yb^{3+}: $^2F_{5/2}$ level populations can be reached from Nd^{3+} ions to Yb^{3+} ions through phonon-assisted energy transfer (PAET) [34,35]. Furthermore, the PAET process between Er^{3+} and Yb^{3+} can be viewed as a $^4I_{13/2}\rightarrow^4I_{15/2}$ transition, because the emission of Er^{3+} can also be found in the fluorescence spectrum of tri-doped Ba_2LuF_7 nanoparticles. In addition, the emission of Ba_2LuF_7: Yb^{3+}/Nd^{3+}/Er^{3+} (1052 nm and 1321 nm) in the near-infrared region matches the $^4F_{3/2}\rightarrow^4F_{11/2}$ and $^4F_{3/2}\rightarrow^4I_{13/2}$ transitions of Nd^{3+} ions [36,37]. Finally, the transitions of Yb^{3+}, Nd^{3+}, and Er^{3+} marked in Figure 2c are distributed in the NIR region of NIR-I, NIR-II, and NIR-III.

In addition, it was found that both green and red upconversion emission intensities were sharply increased by enhancing the excitation optical power, and that the slopes of the emission peaks obtained by linear fitting were 1.213, 1.166, and 1.160, respectively.

Figure S3a,b show that these transitions correspond to two photons, which is beneficial in reducing the energy loss caused by scattered light, thus facilitating fluorescence imaging [38]. Similarly, the same pattern applies to launches located at NIR-I and NIR-II. Notably, the coating of a Ba_2LaF_7 shell onto Ba_2LuF_7: Yb^{3+}/Nd^{3+}/Er^{3+} nanoparticles enhanced the upconversion luminescence intensity 3.5-fold (Figure S3c,d). These results suggest that Ba_2LaF_7 shell passivation can effectively mitigate the quenching of trapped upconversion luminescence on nanocrystal surfaces.

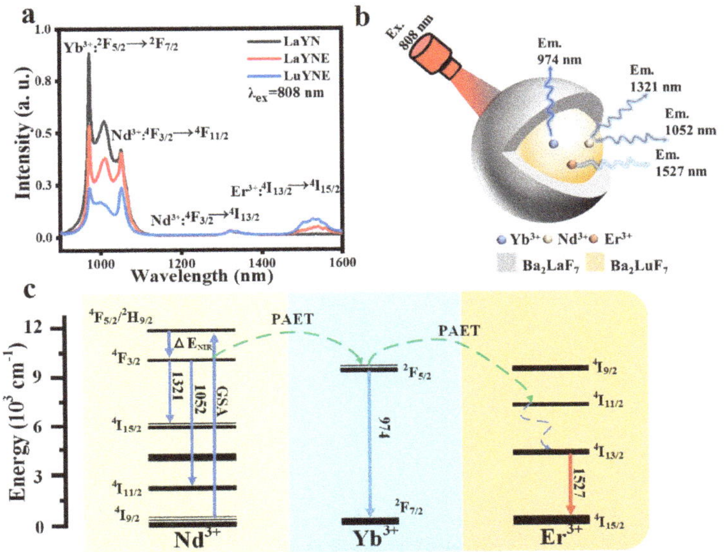

Figure 2. (**a**) The emission spectra of Yb^{3+}/Nd^{3+} and $Yb^{3+}/Er^{3+}/Nd^{3+}$ codoped Ba_2LaF_7 and Ba_2LuF_7 nanoparticles, and the excitation light source wavelength is 808 nm. (**b**) Schematic illustration of the luminescent nanoparticles excited by 808 nm laser. (**c**) General energy level diagram of Er^{3+}, Yb^{3+}, and Nd^{3+} ions.

3.4. Sensitivity of Ratiometric Thermometry for Nanoparticles

The upconversion emission spectra of nanoparticles prepared in this study were fitted as temperature-dependent functions to verify their feasibility as optical thermometers under the premise of temperature control in the 308–528 K range. As disclosed in Figure 3a, the upconversion emission intensity of Ba_2LuF_7: $Yb^{3+}/Nd^{3+}/Er^{3+}$ nanoparticles depends on temperature. Figure 3a specifically shows that the upconversion emission intensity corresponding to the Nd^{3+} ion shows an opposite trend with the increase in temperature, which is triggered by the thermal quenching effect, according to common reports. However, from the integrated upconversion intensities spectral curve at different room temperatures (Figure 3b), the emission intensity of the $^2F_{5/2}\rightarrow^2F_{7/2}$ transition increases slowly compared with that of the $^4F_{3/2}\rightarrow^4F_{11/2}$ transition due to the phonon auxiliary heat of the Nd^{3+} phonon. After it is verified that the FIR values of the two kinds of emissions depend on temperature, it can be predicted that the nanoparticles prepared in this work have the potential to be used in optical temperature sensors. Compared with the thermal coupling energy levels of 540 and 521 nm in Figure S3, the corresponding energy level emissions of 974 and 1052 nm also accord with the Boltzmann factor to control the thermal balance, which can be described as the following:

$$FIR = \frac{I_{974}}{I_{1052}} = \text{Cexp}\left(\frac{-\Delta E}{kT}\right) \quad (1)$$

K and *T* correspond to the Boltzmann constant and Kelvin scale temperature, respectively; *C*, as a constant, is related to the choice of the main material; **Δ***E* is determined by the energy gap between the $^4F_{3/2}$ and $^2F_{5/2}$ states. As disclosed in Figure 3b, the relative value of I_{1052} weakened steadily. Meanwhile, the decline of I_{1052} exceeds that of I_{974}. I_{974}/I_{1052}

reveals an escalating trend from 0.88775 to 2.12769, because the downward trends in I_{974} and I_{1052} were different.

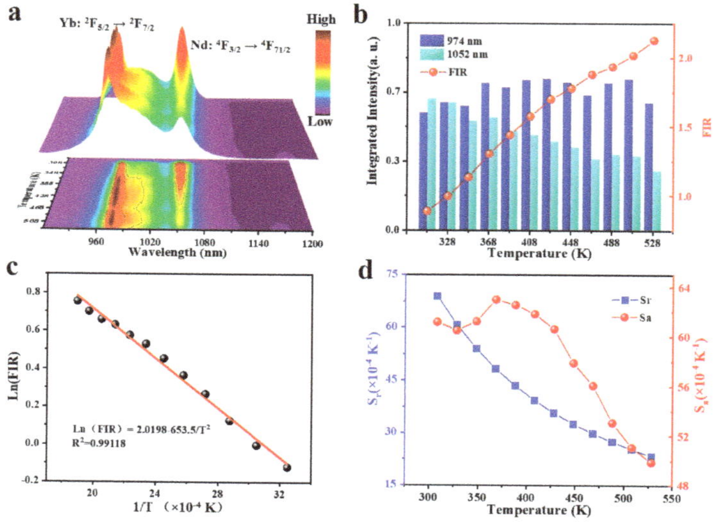

Figure 3. (**a**) Photoluminescence spectra curve of the Ba_2LuF_7: $Yb^{3+}/Nd^{3+}/Er^{3+}$ nanoparticles covering 308–528 K temperature range. Integrated upconversion intensities at 974 nm and 1052 nm. (**b**) Temperature dependence of the fluorescence ratio values of thermally coupled energy levels. (**c**) Ln(FIR) is an inverse absolute function of absolute temperature. (**d**) The temperature function of the thermal coupling level of nanomaterials based on the fitting of Sa and Sr values.

The absolute temperature ($1/T$) and $\ln(I_{974}/I_{1052})$ were linearly fitted according to Equation (1). This is an inverse curve with a regression coefficient (R^2) of 0.99829, and all points could pass through the fitting line, as disclosed in Figure 3c. The relevant parameters of the fitted curve are −0.06535 and 2.0198, respectively, corresponding to the slope and intercept. The fitting curve of $\mathrm{Ln}(I_{974}/I_{1052})$ and $1/T$ has a good linear trend in the temperature range of 308 K to 528 K.

As an important index to measure optical temperature sensitivity, S_a represents absolute sensitivity and S_r represents relative sensitivity. The value of S_a can be calculated by the following formula:

$$S_a = \left| \frac{dFIR}{dT} \right| = FIR\left(\frac{\Delta E}{kT^2} \right) \tag{2}$$

Figure 3d shows the function curve of S_a and S_r from 308 to 548 K concerning temperature. Since the slope and intercept of the fitted line were obtained in Figure 3c, the values of S_r and S_a can be obtained only by substituting them into Equations (3) and (2). As disclosed in the figure, due to the influence of the auxiliary heat of the Nd^{3+} ion phonon, the greatest S_a first occurs when the curve is at 368 K, which is 0.50% K^{-1}, and then it gradually decreases as the temperature goes up. When the temperature goes up, S_r decreases gradually. At this point, the greatest value is 0.63% K^{-1} at 308 K.

As an important parameter to measure the performance of optical thermometers, the relative temperature can be calculated by the following formula:

$$S_r = \left| \frac{1}{\mathrm{FIR}} \frac{dT}{\mathrm{dFIR}} \right| \times 100\% \tag{3}$$

Because a certain amount of Nd^{3+} ions were added into the prepared Yb^{3+}/Er^{3+} nanocrystals in this work, a new emission peak in the near-infrared region of the excitation spectrum could be observed. The nanocrystals were exposed to an 808 nm light source. For instance, the peak of the spectrum at 1321 nm is the energy level transition of Nd^{3+}: $^4I_{3/2}\rightarrow{}^4I_{15/2}$. It can be understood from the revelation in Figure 4a that the upconversion luminescence intensity declines with the rise in temperature, within the range of 308 ~ 548 K. Similarly, due to the phonon-assisted heat of Er^{3+} phonons, the emission intensity of the $^4I_{13/4}\rightarrow{}^4I_{15/2}$ transition fluctuates with the rise and fall of the experimental temperature. Figure 4b shows a comparison of the luminescence intensity from 308 to 548 K with temperature changes at 1321 nm and 1527 nm wavelengths. As mentioned above, even though the luminescence intensity corresponding to the 1321 nm level declines with the rise in temperature, the luminescence intensity corresponding to the 1527 nm level decays faster under the same conditions. This will cause the FIR value of I_{1321}/I_{1527} to decline when the temperature rises. It has the greatest FIR of 0.31652 at 308 K and the least FIR of 0.22582 at 548 K.

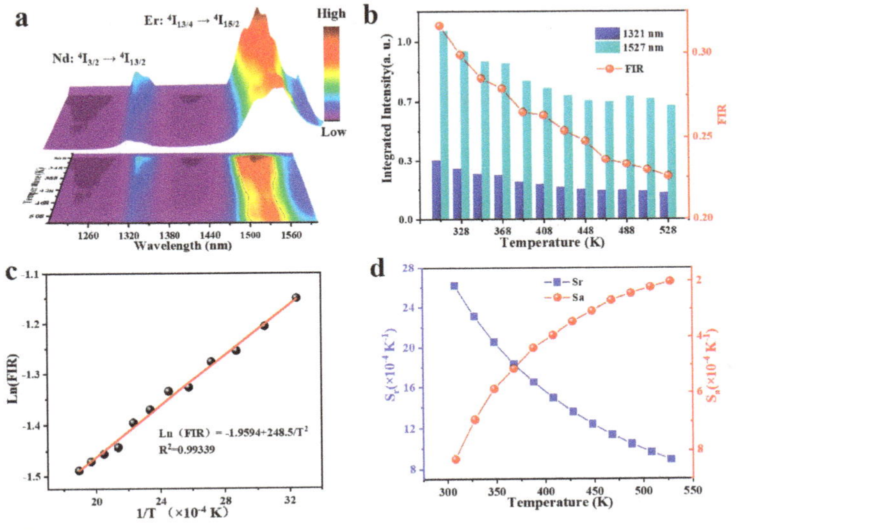

Figure 4. (**a**) Photoluminescence emission spectra curve of the Ba_2LuF_7: $Nd^{3+}/Yb^{3+}/Er^{3+}$ nanoparticles covering 308–528 K temperature range. Integrated upconversion intensities at 1321 nm and 1527 nm. (**b**) Temperature dependence of the fluorescence ratio values of thermally coupled energy levels. Dependence of FIR values of the (**b**) thermally coupled levels on temperature. (**c**) Ln(FIR) as an inverse absolute function of inverse absolute temperature. (**d**) The temperature function of the thermal coupling level of nanomaterials based on the fitting of S_a and S_r values. S_a and S_r values are based on the (**d**) thermally coupled levels as a function of temperature.

As shown in Figure 4c, Ln(FIR) was linearly fitted to $1/T$ by Equation (1) to obtain a line with $R^2 = 0.99339$. According to the fitting results, the calculated Ln(FIR) value coincides with the fitting line. The relevant parameters of the fitted curve are 0.2465 and −1.9594, respectively. Figure 4d reveals the curve composed of the values of S_a and S_r calculated by substituting them into Equations (2) and (3). In the range of 308 K to 548 K, S_a rose with the rise in temperature, and the greatest value was 0.08% K^{-1}. The variation trend of the S_r value was opposite to that of S_a, and the greatest value was 0.262% K^{-1}.

3.5. Fluorescence Imaging of Nanoparticles in Biological Windows

According to the above data, nanoparticles have the potential to be applied in multimode fluorescence temperature measurement, which provides the possibility for the universality of non-contact tissue temperature measurement. As shown in Figure S4, the non-contact optical temperature measurement system simulates biological tissues at different depths by controlling lipid thickness. Nanoparticles are injected into biological tissues. The light received under the excitation of the 808 nm external excitation light source more easily penetrates the NIR region of biological tissues than the visible light spectrometer. The NIR region used by different temperature measurement modes in this work can be found in Figure S5. These thermometric models are based on the energy transfer process between $Yb^{3+}/Nd^{3+}/Er^{3+}$. Nd^{3+} ions in the first region and the second region are used as the selector switch to determine the fluorescence ratio temperature measurement for narrowband separation and broad-band separation, respectively. The fluorescence ratio thermometry of mode-two width separation can detect deeper tissues thanks to the excellent penetration of NIR region 2, and the fluorescence ratio thermometry of mode-one narrow separation has higher sensitivity.

NIR window luminescence has low water absorption efficiency and good tissue penetration, which is often used in biological tissue imaging [39–42]. Taking the two emissions of region 1 and region 2 in mode 1 as an example (Figure 5a), it can be noticed that core–shell nanoparticles in the darkroom under the NIR source glow brightly and clearly in the field of view of the infrared camera (Figure 5b). As shown in Figure 5c, chicken tissues with thicknesses of 2 mm, 4 mm, and 6 mm were selected for imaging. In the image of the 2 mm tissue there are shaded areas, which are attached bone sheets on the back of the biological tissue. This indicates that the NIR emission of nanocrystals has different transmittance to different tissue components, which makes it possible to image foreign bodies in tissues. In 4 mm and 6 mm tissue imaging, more obvious cartilage tissue and ribs can be observed, which further confirms the imaging ability of nanocrystals. In addition, the addition of 12 mm human pinky finger imaging still allows relatively clear observation of bones and joints, indicating the potential of nanocrystals for deep tissue imaging. Furthermore, image quality improvement combined with algorithm image processing can further improve image clarity to achieve auxiliary tumor tissue and blood vessel imaging, which reflects the potential of nanocrystals in fine tissue imaging. Hence, the nanoparticles are injected into mice after being biologically modified to help image specific tissues and organs through the mice's circulatory system. As for multi-mode fluorescence ratio temperature measurement, the receiving window can be selected according to the depth of the desired imaging location due to the stronger tissue penetration of the NIR second region window. Taken together, these results finally show that our nanocrystals have some possibility for imaging in biological windows.

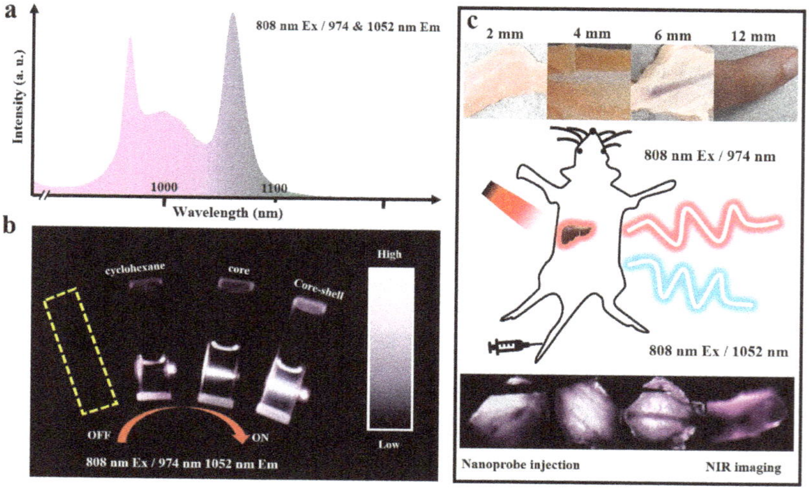

Figure 5. (**a**) The selected NIR imaging interval. (**b**) NIR luminescence of nanoparticles. (**c**) Schematic of biological tissue imaging of nanoparticles.

4. Conclusions

All in all, we worked on the non-contact multifunctional biological applications of lanthanide-doped nanoparticles based on NIR luminescence. On this basis, the advantages of FIR optical thermometry technology were shown, and the maximum S_r is close to 0.63% K^{-1} in the range of 308–538 K. The FIR temperature measurement mode can be selected according to the tissue depth of the temperature measurement site. The flexible selection broadens the application scenario and gives broad application prospects in the field of optical thermometry. At the same time, the NIR imaging of the material reveals that the temperature-assisted imaging of the material works well in practical applications. In the end, the lanthanide-doped nanoparticles designed in this paper have potential applications in NIR luminescence-assisted imaging and multi-mode optical temperature measurement of biological tissues.

Supplementary Materials: The following supporting information can be downloaded at: https://www.mdpi.com/article/10.3390/nano13010219/s1, Table S1. Comparison of optical properties of some nanomaterials and their applications in proportional nanothermometers; Figure S1. (a) UV absorption spectra of nanoparticles in cyclohexane. (b) Fourier infrared spectroscopy of nanoparticles; Figure S2. (a,b) XPS survey spectra of the Ba2LuF7: Yb3+/Nd3+/Er3+ and Ba2LuF7: Yb3+/Nd3+/Er3+@Ba2LaF7 nanoparticles, respectively. And deconvoluted spectra of (c) Ba 3d, (d) La 3d, (e) Lu 4d, and (f) F 1s region of Ba2LuF7@Ba2LaF7; Figure S3. (a) Upconversion luminescence spectra in visible regions of nanocrystals excited at different power densities of 980 nm. (b) The relationship between emission intensity and excitation power. (c,d) Comparison of upconversion luminescence intensity between core and core-shell nanoparticles; Figure S4. Diagram of the non-contact temperature measuring system; Figure S5. Schematic diagram of multi-mode temperature measurement principle.

Author Contributions: Conceptualization, methodology, software, validation, formal analysis, investigation, visualization, writing—original draft preparation, H.L.; writing—reviewing and editing, G.B., E.H., Y.L., H.X., S.X. and L.C.; project administration, resources, supervision, G.B. All authors have read and agreed to the published version of the manuscript.

Funding: This work was funded by the Zhejiang Provincial Natural Science Foundation of China (LZ21E020004) and the National Natural Science Foundation of China (6217032304).

Acknowledgments: This work was supported by the Zhejiang Provincial Natural Science Foundation of China (LZ21E020004) and the National Natural Science Foundation of China (6217032304).

Conflicts of Interest: The authors declare no conflict of interest.

References

1. Zheng, B.; Fan, J.; Chen, B.; Qin, X.; Wang, J.; Wang, F.; Deng, R.; Liu, X. Rare-earth doping in nanostructured inorganic materials. *Chem. Rev.* **2022**, *122*, 5519–5603. [CrossRef]
2. An, R.; Liang, Y.; Deng, R.; Lei, P.; Zhang, H. Hollow nanoparticles synthesized via Ostwald ripening and their upconversion luminescence-mediated Boltzmann thermometry over a wide temperature range. *Light Sci. Appl.* **2022**, *11*, 217. [CrossRef] [PubMed]
3. Otari, S.V.; Kumar, M.; Anwar, M.Z.; Thorat, N.D.; Patel, S.K.; Lee, D.; Lee, J.H.; Lee, J.K.; Kang, Y.C.; Zhang, L. Rapid synthesis and decoration of reduced graphene oxide with gold nanoparticles by thermostable peptides for memory device and photothermal applications. *Sci. Rep.* **2017**, *7*, 10980. [CrossRef] [PubMed]
4. Zhou, B.; Shi, B.; Jin, D.; Liu, X. Controlling upconversion nanocrystals for emerging applications. *Nat. Nanotechnol.* **2015**, *10*, 924–936. [CrossRef] [PubMed]
5. Du, P.; Tang, J.; Li, W.; Luo, L. Exploiting the diverse photoluminescence behaviors of NaLuF4: xEu^{3+} nanoparticles and g-C3N4 to realize versatile applications in white light-emitting diode and optical thermometer. *Chem. Eng. J.* **2021**, *406*, 127165. [CrossRef]
6. Jia, M.; Chen, X.; Sun, R.; Wu, D.; Li, X.; Shi, Z.; Chen, G.; Shan, C. Lanthanide-based ratiometric luminescence nanothermometry. *Nano Res.* **2022**. [CrossRef]
7. Sun, Z.; Jia, M.; Wei, Y.; Cheng, J.; Sheng, T.; Fu, Z. Constructing new thermally coupled levels based on different emitting centers for high sensitive optical thermometer. *Chem. Eng. J.* **2020**, *381*, 122654. [CrossRef]
8. Xue, J.; Yu, Z.; Noh, H.M.; Lee, B.R.; Choi, B.C.; Park, S.H.; Jeong, J.H.; Du, P.; Song, M. Designing multi-mode optical thermometers via the thermochromic LaNbO4: Bi^{3+}/Ln^{3+} (Ln = Eu, Tb, Dy, Sm) phosphors. *Chem. Eng. J.* **2021**, *415*, 128977. [CrossRef]
9. Gupta, C.; Pauzauskie, P.J. Semiconductor yields sensitive thermometry. *Nat. Photonics* **2022**, *16*, 407–408. [CrossRef]

10. Ansari, A.A.; Parchur, A.K.; Nazeeruddin, M.K.; Tavakoli, M.M. Coord. Luminescent lanthanide nanocomposites in thermometry: Chemistry of dopant ions and host matrices. *Chem. Rev.* **2021**, *444*, 214040.
11. Kondaveeti, S.; Park, G.D.; Shanmugam, R.; Pagolu, R.; Patel, S.K.; Bisht, A.; Kim, D.R.; Kang, Y.C.; Lee, J.K. Investigating the role of metals loaded on nitrogen-doped carbon-nanotube electrodes in electroenzymatic alcohol dehydrogenation. *Appl. Catal. B* **2022**, *307*, 121195. [CrossRef]
12. Chen, X.; Liu, S.; Huang, K.; Nie, J.; Kang, R.; Tian, X.; Zhang, S.; Li, Y.; Qiu, J. Cr^{4+} activated NIR-NIR multi-mode luminescent nanothermometer for double biological windows. *J. Chem. Eng. J.* **2020**, *396*, 125201. [CrossRef]
13. Wang, Z.; Jia, M.; Zhang, M.; Jin, X.; Xu, H.; Fu, Z. Trimodal Ratiometric Luminescent Thermometer Covering Three Near-Infrared Transparency Windows. *Inorg. Chem.* **2021**, *60*, 14944–14951. [CrossRef]
14. Xie, Y.; Song, Y.; Sun, G.; Hu, P.; Bednarkiewicz, A.; Sun, L. Lanthanide-doped heterostructured nanocomposites toward advanced optical anti-counterfeiting and information storage. *Light Sci. Appl.* **2022**, *11*, 150. [CrossRef]
15. Wang, F.; Deng, R.; Wang, J.; Wang, Q.; Han, Y.; Zhu, H.; Chen, X.; Liu, X. Tuning upconversion through energy migration in core–shell nanoparticles. *Nat. Mater.* **2011**, *10*, 968–973. [CrossRef] [PubMed]
16. Li, H.; Wang, X.; Ohulchanskyy, T.Y.; Chen, G. Lanthanide-doped near-infrared nanoparticles for biophotonics. *Adv. Mater.* **2021**, *33*, 2000678. [CrossRef]
17. Min, X.; Sun, Y.; Kong, L.; Guan, M.; Fang, M.; Liu, Y.G.; Wu, X.; Huang, Z. Novel pyrochlore-type $La_2Zr_2O_7$: Eu^{3+} red phosphors: Synthesis, structural, luminescence properties and theoretical calculation. *Dyes Pigm.* **2018**, *157*, 47–54. [CrossRef]
18. Zheng, S.; Zhang, Z.; Qu, Y.; Zhang, X.; Guo, H.; Shi, X.; Cai, M.; Cao, C.; Hu, Z.; Liu, H.; et al. Radiopharmaceuticals and Fluorescein Sodium Mediated Triple-Modality Molecular Imaging Allows Precise Image-Guided Tumor Surgery. *Adv. Sci.* **2019**, *6*, 1900159. [CrossRef]
19. Nexha, A.; Carvajal, J.J.; Pujol, M.C.; Díaz, F.; Aguiló, M. Lanthanide doped luminescence nanothermometers in the biological windows: Strategies and applications. *Nanoscale* **2021**, *13*, 7913–7987. [CrossRef]
20. Yang, J.; He, S.; Hu, Z.; Zhang, Z.; Cao, C.; Cheng, Z.; Fang, C.; Tian, J. In vivo multifunctional fluorescence imaging using liposome-coated lanthanide nanoparticles in near-infrared-II/IIa/IIb windows. *Nano Today* **2021**, *38*, 101120. [CrossRef]
21. Zhou, T.; Cheng, Q.; Zhang, L.; Zhang, D.; Li, L.; Jiang, T.; Huang, L.; Xu, H.; Hu, M.; Jing, S. Ferrocene-functionalized core–shell lanthanide-doped upconversion nanoparticles: NIR light promoted chemodynamic therapy and luminescence imaging of solid tumors. *Chem. Eng. J.* **2022**, *438*, 135637. [CrossRef]
22. Fan, Y.; Liu, L.; Zhang, F. Exploiting lanthanide-doped upconversion nanoparticles with core/shell structures. *Nano Today* **2019**, *25*, 68–84. [CrossRef]
23. Ding, M.; Dong, B.; Lu, Y.; Yang, X.; Yuan, Y.; Bai, W.; Wu, S.; Ji, Z.; Lu, C.; Zhang, K.; et al. Energy manipulation in lanthanide-doped core–shell nanoparticles for tunable dual-mode luminescence toward advanced anti-counterfeiting. *Adv. Mater.* **2020**, *32*, 2002121. [CrossRef] [PubMed]
24. Kaczmarek, A.M.; Suta, M.; Rijckaert, H.; Van Swieten, T.P.; Van Driessche, I.; Kaczmarek, M.K.; Meijerink, A. High temperature (nano) thermometers based on $LiLuF_4$: Er^{3+}, Yb^{3+} nano-and microcrystals. Confounded results for core–shell nanocrystals. *J. Mater. Chem. C* **2021**, *9*, 3589–3600. [CrossRef]
25. Geitenbeek, R.G.; Salzmann, B.B.; Nieuwelink, A.E.; Meijerink, A.; Weckhuysen, B.M. Operando nanoscale sensors in catalysis: All eyes on catalyst particles. *Chem. Eng. Sci.* **2019**, *198*, 235–240. [CrossRef]
26. Zhou, Y.; Cheng, Y.; Huang, Q.; Xu, J.; Lin, H.; Wang, Y. Abnormal thermally enhanced upconversion luminescence of lanthanide-doped phosphors: Proposed mechanisms and potential applications. *J. Mater. Chem. C* **2021**, *9*, 2220–2230. [CrossRef]
27. Zhao, H.; Vomiero, A.; Rosei, F. Tailoring the heterostructure of colloidal quantum dots for ratiometric optical nanothermometry. *Small* **2020**, *16*, 2000804. [CrossRef]
28. Haro-González, P.; Martínez-Maestro, L.; Martín, I.R.; García-Solé, J.; Jaque, D. High-sensitivity fluorescence lifetime thermal sensing based on CdTe quantum dots. *Small* **2012**, *8*, 2652–2658. [CrossRef]
29. Lei, P.; An, R.; Li, C.; Feng, J.; Zhang, H. Lanthanide-doped bismuth-based fluoride nanoparticles: Controlled synthesis and ratiometric temperature sensing. *CrystEngComm* **2020**, *22*, 3432–3438. [CrossRef]
30. Zhao, Y.; Bai, G.; Huang, Y.; Liu, Y.; Peng, D.; Chen, L.; Xu, S. Stimuli responsive lanthanide ions doped layered piezophotonic microcrystals for optical multifunctional sensing applications. *Nano Energy* **2021**, *87*, 106177. [CrossRef]
31. Pu, L.; Wang, Y.; Zhao, J.; Jin, M.; Li, L.; Li, P.; Wang, Z.; Guo, C.; Suo, H. Multi-mode ratiometric thermometry using thermo-intensified NIR emission. *Chem. Eng. J.* **2022**, *449*, 137890. [CrossRef]
32. Maurizio, S.L.; Tessitore, G.; Krämer, K.W.; Capobianco, J.A. $BaYF_5$: Yb^{3+}, Tm^{3+} upconverting nanoparticles with improved population of the visible and near-infrared emitting states: Implications for bioimaging. *ACS Appl. Nano Mater.* **2021**, *4*, 5301–5308. [CrossRef]
33. Yu, X.; Lu, K.; Chen, L. Synergistic effects of oxygen vacancies and the chelation of tetracycline with metallic ions for enhanced degradation of tetracycline over photocatalysts $La_{2-x}K_xNiMnO_6$. *CrystEngComm* **2021**, *23*, 1455–1471. [CrossRef]
34. Kolesnikov, I.E.; Afanaseva, E.V.; Kurochkin, M.A.; Vaishlia, E.I.; Kalinichev, A.A. Upconverting NIR-to-NIR $LuVO_4$: Nd^{3+}/Yb^{3+} Nanophosphors for High-Sensitivity Optical Thermometry. *ACS Appl. Mater. Interfaces* **2022**, *14*, 1757–1764. [CrossRef] [PubMed]
35. Yu, D.; Yu, T.; van Bunningen, A.J.; Zhang, Q.; Meijerink, A.; Rabouw, F.T. Understanding and tuning blue-to-near-infrared photon cutting by the Tm^{3+}/Yb^{3+} couple. *Light Sci. Appl.* **2020**, *9*, 107. [CrossRef] [PubMed]

36. Marciniak, L.; Piotrowski, W.M.; Drozd, M.; Kinzhybalo, V.; Bednarkiewicz, A.; Dramicanin, M. Activators Lattice Migration Strategy Customized for Tunable Luminescence of Ce^{3+} Doped β-Ca_3 $(PO_4)_2$. *Adv. Opt. Mater.* **2022**, *10*, 2102856. [CrossRef]
37. Yuliantini, L.; Djamal, M.; Hidayat, R.; Boonin, K.; Kaewkhao, J.; Yasaka, P. Luminescence and Judd-Ofelt analysis of Nd^{3+} ion doped oxyfluoride boro-tellurite glass for near-infrared laser application. *Mater. Today Proc.* **2021**, *43*, 2655–2662. [CrossRef]
38. Janiak, F.K.; Bartel, P.; Bale, M.R.; Yoshimatsu, T.; Komulainen, E.; Zhou, M.; Staras, K.; Prieto-Godino, L.L.; Euler, T.; Maravall, M.; et al. Non-telecentric two-photon microscopy for 3D random access mesoscale imaging. *Nat. Commun.* **2022**, *13*, 544. [CrossRef]
39. Lin, H.; Lin, Z.; Zheng, K.; Wang, C.; Lin, L.; Chen, J.; Song, J. Near-Infrared-II Nanomaterials for Fluorescence Imaging and Photodynamic Therapy. *Adv. Opt. Mater.* **2021**, *9*, 2002177. [CrossRef]
40. Ding, F.; Feng, J.; Zhang, X.; Sun, J.; Fan, C.; Ge, Z. Responsive optical probes for deep-tissue imaging: Photoacoustics and second near-infrared fluorescence. *Adv. Drug Deliv. Rev.* **2021**, *173*, 141–163. [CrossRef]
41. Pei, P.; Chen, Y.; Sun, C.; Fan, Y.; Yang, Y.; Liu, X.; Lu, L.; Zhao, M.; Zhang, H.; Zhao, D.; et al. X-ray-activated persistent luminescence nanomaterials for NIR-II imaging. *Nat. Nanotechnol.* **2021**, *16*, 1011–1018. [CrossRef] [PubMed]
42. Lv, Z.; Jin, L.; Cao, Y.; Zhang, H.; Xue, D.; Yin, N.; Zhang, T.; Wang, Y.; Liu, J.; Liu, X.; et al. A nanotheranostic agent based on Nd^{3+}-doped YVO_4 with blood-brain-barrier permeability for NIR-II fluorescence imaging/magnetic resonance imaging and boosted sonodynamic therapy of orthotopic glioma. *Light Sci. Appl.* **2022**, *11*, 1–10. [CrossRef] [PubMed]

Disclaimer/Publisher's Note: The statements, opinions and data contained in all publications are solely those of the individual author(s) and contributor(s) and not of MDPI and/or the editor(s). MDPI and/or the editor(s) disclaim responsibility for any injury to people or property resulting from any ideas, methods, instructions or products referred to in the content.

nanomaterials

MDPI

Article

Carbon Nanodots as Electron Transport Materials in Organic Light Emitting Diodes and Solar Cells

Zoi Georgiopoulou [1,2], Apostolis Verykios [1], Kalliopi Ladomenou [3], Katerina Maskanaki [4], Georgios Chatzigiannakis [1,2], Konstantina-Kalliopi Armadorou [1], Leonidas C. Palilis [5], Alexander Chroneos [6,7,*], Evangelos K. Evangelou [4], Spiros Gardelis [2], Abd. Rashid bin Mohd Yusoff [8], Athanassios G. Coutsolelos [9,*], Konstantinos Aidinis [10,11,*], Maria Vasilopoulou [1,*] and Anastasia Soultati [1,*]

1 Institute of Nanoscience and Nanotechnology (INN), National Center for Scientific Research Demokritos, Agia Paraskevi, 15341 Athens, Greece
2 Solid State Physics Section, Physics Department, National and Kapodistrian University of Athens, Panepistimioupolis, Zografos, 15784 Athens, Greece
3 Department of Chemistry, International Hellenic University, 65404 Kavala, Greece
4 Department of Physics, University of Ioannina, 45110 Ioannina, Greece
5 Department of Physics, University of Patras, Rio, 26504 Patra, Greece
6 Department of Electrical and Computer Engineering, University of Thessaly, 38221 Volos, Greece
7 Department of Materials, Imperial College, London SW7 2AZ, UK
8 Department of Chemical Engineering, Pohang University of Science and Technology (POSTECH), Pohang 37673, Gyeongbuk, Republic of Korea
9 Laboratory of Bioinorganic Chemistry, Department of Chemistry, University of Crete, Voutes Campus, Heraklion, 71003 Crete, Greece
10 Department of Electrical and Computer Engineering, Ajman University, Ajman P.O. Box 346, United Arab Emirates
11 Center of Medical and Bio-allied Health Sciences Research, Ajman P.O. Box 388, United Arab Emirates
* Correspondence: alexander.chroneos@imperial.ac.uk (A.C.); acoutsol@uoc.gr (A.G.C.); k.aidinis@ajman.ac.ae (K.A.); m.vasilopoulou@inn.demokritos.gr (M.V.); a.soultati@inn.demokritos.gr (A.S.)

Abstract: Charge injection and transport interlayers play a crucial role in many classes of optoelectronics, including organic and perovskite ones. Here, we demonstrate the beneficial role of carbon nanodots, both pristine and nitrogen-functionalized, as electron transport materials in organic light emitting diodes (OLEDs) and organic solar cells (OSCs). Pristine (referred to as C-dots) and nitrogen-functionalized (referred to as NC-dots) carbon dots are systematically studied regarding their properties by using cyclic voltammetry, Fourier-transform infrared (FTIR) and UV–Vis absorption spectroscopy in order to reveal their energetic alignment and possible interaction with the organic semiconductor's emissive layer. Atomic force microscopy unravels the ultra-thin nature of the interlayers. They are next applied as interlayers between an Al metal cathode and a conventional green-yellow copolymer—in particular, (poly[(9,9-dioctylfluorenyl-2,7-diyl)-alt-co-(1,4-benzo-{2,1′,3}-thiadiazole)], F8BT)—used as an emissive layer in fluorescent OLEDs. Electrical measurements indicate that both the C-dot- and NC-dot-based OLED devices present significant improvements in their current and luminescent characteristics, mainly due to a decrease in electron injection barrier. Both C-dots and NC-dots are also used as cathode interfacial layers in OSCs with an inverted architecture. An increase of nearly 10% in power conversion efficiency (PCE) for the devices using the C-dots and NC-dots compared to the reference one is achieved. The application of low-cost solution-processed materials in OLEDs and OSCs may contribute to their wide implementation in large-area applications.

Keywords: organic light emitting diodes; organic solar cells; carbon nanodots; surface functionalization; electron transport materials

1. Introduction

In 1987, Tang and Van Slyke introduced the first electroluminescent device based on organic semiconductors, the so-called organic light emitting diode (OLED) [1]. After three decades, they are considered a mature technology and have already achieved successful market entry. However, OLEDs still attract research interest because of their potential as a promising technology for solid state lighting and flat panel displays [2], considering their low-cost fabrication, flexibility and large viewing angle. The balance of charge injection and the transport of holes and electrons through the organic semiconductor plays a vital role regarding the device's (OLED and OSC) efficiency [3]. One of the key issues for the efficient design of organic optoelectronic devices is the understanding of the energy-level alignment at the metal contact/organic semiconductor interfaces. Proper matching of the electrode Fermi level to the energy level of the charge transport states of the organic semiconductor is necessary to obtain efficient and balanced charge injection in these organic electronic devices [4]. This is why, in highly efficient devices, anode and cathode interfacial layers are necessary to enhance the charge exchange between metal contacts and organic molecules.

A typical structure of an OLED and/or OSC is based on a multilayer architecture [5]. The emissive and/or photoactive layer and the various interlayers (hole injection layers (HILs), hole transport layers (HTLs), electron blocking layers (EBLs), hole blocking layers (HBLs), electron transport layers (ETLs) and electron injection layers (EILs)) are sandwiched between the two electrodes (anode and cathode). It is well known that in organic electroluminescent devices, the mobility of holes is larger than that of electrons and the hole injection barrier is lower than the electron injection barrier [6]. Such devices have extra holes in the emission region, and electrons are more or less consumed before reaching the emissive layer, which decreases the device's efficiency. An effective approach is to modify the cathode, aiming to reduce the energy barrier and increase the charge transport between the metal cathode and the emissive and/or photoactive layer. This can be accomplished by inserting suitable interfacial layers (ETL, HBL, EIL) [7,8]. A large number of organic and inorganic materials have been applied as ETLs in OLEDs and OSCs to lower the electron injection barrier and achieve charge carrier balance. They include transition metal oxides with intrinsic n-type conductivity, such as zinc oxide (ZnO) [9–12] and tin dioxide (SnO_2) [13,14], and mostly organic molecules bearing suitable functional groups, such as perylene [15,16] and porphyrin [17–19] compounds, polymeric oxadiazoles, metal chelates, azole-based materials and triazine [20–25]. However, despite their great success, many of these materials require vacuum deposition methods, which are incompatible with large-area devices for low-cost applications. The development of solution-processable materials of potentially low cost and with facile synthesis and deposition methods still represents a topic of intense research interest in OLEDs and OSCs.

Carbon nanodots are advantageous in terms of the abundant and sustainable precursor materials and facile surface functionalization to promote their beneficial properties. The facile and low-cost preparation/modification of these materials motivated us to further explore their use as ETLs in high-performance green emitting OLEDs and inverted OSCs. We synthesized pristine and nitrogen-functionalized carbon dots (described hereafter as C-dots and NC-dots, respectively) and unraveled their effects on the OLED and OSC performance, which paves the way for the further exploration of not only organic but also perovskite optoelectronics.

2. Materials and Methods

2.1. OLED Device Fabrication Procedure

The first and most fundamental step before performing the layers' deposition was the proper cleaning of the glass substrates. We utilized coated glass substrates of indium-tin oxide (ITO) purchased from Sigma-Aldrich (Athens, Greece), with sheet resistance of 15–25 Ω/sq, which served as the transparent anode electrode. ITO-coated glasses were placed into an ultrasonic cleaner, where they remained for ten minutes in each bath: first in deionized water, then in acetone and finally in isopropyl alcohol (IPA). Moreover, all samples were dried with N_2 gas before each bath. Then, the ITO substrates were subjected to UV–ozone treatment for 20 min. Next, we performed the deposition of hole injection layers (HTLs) using the commercially available solution PEDOT:PSS (poly(3,4 - ethylenedioxythiophene)−poly(styrenesulfonate)) with 1.3 wt % dispersion in H_2O, from Sigma-Aldrich. At first, it was passed through a 0.45 μm pore diameter polyvinylidene fluoride (PVDF) filter and then was spin-coated on ITO-coated glass substrates at 6000 rpm for 40 s, forming a 40 nm thick layer. The substrates were annealed at 110 °C for at least 30 min on a hotplate. A green-yellow copolymer, named F8BT (poly[(9,9-dioctylfluorenyl-2,7-diyl)-alt-co-(1,4-benzo-{2,1′,3}-thiadiazole)), purchased from the American Dyes Source, Quebec, Canada, (ADS 233 YE), used as the emissive layer. F8BT was filtered through a 0.22 μm pore diameter PTFE filter and was deposited on top of the PEDOT:PSS. The deposition was carried out by spin-coating at 1200 rpm for 40 s from a 10 mg mL^{-1} solution in chloroform, forming a 80 nm thick layer. The substrates were annealed at 85 °C for 10 min on a hotplate. Moreover, carbon dots were employed as EILs. In particular, 50 μL of carbon dots (C-dots) and nitrogen-doped carbon dots (NC-dots) was spin-coated at 2000 rpm for 40 s from a 0.5 mg mL^{-1} solution in methanol atop the emissive layer. Finally, the OLED devices were completed with the deposition of a 150 nm aluminum layer through thermal evaporation in order to serve as the cathode electrode. The active surface of each diode was set at 12.56 mm^2.

2.2. OSC Device Fabrication Procedure

Glass/ITO substrates were cleaned as already mentioned and used as the transparent cathode electrode for the fabrication of the inverted OSCs. Tin oxide purchased from Alfa Aesar was deposited from a water solution (15% colloidal dispersion of tin(vi) oxide in deionized water was diluted in 1:6.5 water) with spin-coating at 3000 rpm for 30 s. Then, the samples were thermally annealed at 150 °C for 30 min. PTB7-Th:PCBM (purchased from Ossila, Sheffield, UK), at 10 mg:15 mg in 1 mL 1,2-dichlorobenzene with the addition of 3% per volume 1,8 diiodoctane (DIO), was spin-coated at 1000 rpm for 90 s on the SnO_2 ETL to serve as the photoactive layer. The fabrication procedure was completed with the deposition of molybdenum oxide (MoO_x) HTL by a hot-wire CVD method [26] and Al anode electrode by thermal evaporation.

2.3. Characterization Methods

Current density–voltage–luminance (I−V−L) measurements were performed using a Keithley 2601A Power Supply Source-Meter (Vector Technologies, Athens, Greece) in voltage mode, with constant increment steps. The instrument was used both for slow (delay time of 1 sec before each measurement point) and fast (delay time of <1 msec) measurements. EL analysis was performed using a calibrated photodiode (BPW34 Si PIN photodiode) (Vector Technologies, Athens, Greece) connected to a Keithley 6500 DMM (Vector Technologies, Athens, Greece). The same instruments were used for the current and luminance–time (I-t and L-t) measurements. In this case, the current measurements through the diodes were acquired at rates of up to 1K sample/sec, while the luminance values were obtained at faster rates, reaching values of up to 1M sample/sec. All measurements were performed at room temperature in a dark shielded probe station to minimize light and EM interference. For the OSCs' electrical characterization in dark and illuminated conditions, a Keithley 2400 source meter unit and a Xe lamp with an AM 1.5G filter were used. TEM measurements were performed using a Philips CM 20 transmission electron microscope (Thermo fisher scientific, United States). UV–Vis absorbance and transmittance spectra were captured by a Perkin Elmer Lambda 40 UV–Vis spectrometer (Vamvakas-Scientific Equipment, Athens, Greece). XRD patterns were recorded using a Smart Lab Rigaku diffractometer (Japan) with Cu Kα radiation. Cyclic voltammetry measurements were performed using a VersaSTAT 4 potentiometer (Megalab SA, Athens, Greece). Fourier-transform infrared (FT-IR) spectroscopy was performed using a Bruker Tensor 27 spectrometer (Interactive, Athens, Greece), with a DTGS detector. The surface morphology of devices was recorded with an NT-MDT AFM system (LaborScience SA, Athens, Greece) in tapping operation mode. Ellipsometry measurements were carried out with a J.A Woolam Inc. M2000F rotating compensator ellipsometer that operated within the 250–1000 nm range, running the WVASE32 software, at an angle of incidence of 75.14°. Steady-state PL measurements were performed with a commercial platform (ARKEO—Cicci Research). In particular, the substrate was illuminated with a diode-pumped solid-state Nd:YVO4 + KTP laser (peak wavelength 532 nm ± 1 nm, optical power 1 mW on a circular spot of 2 mm of diameter 31 mW cm^{-2}) at an inclination of 45°. The fluorescence on the opposite side of the substrate was focused on a bundle of fibers (10 mm in diameter) with an aspheric lens close to the substrate to maximize the PL. The bundle sent the signal to a CCD-based spectrometer. The integration time and the number of averages was maintained to better compare the results. Time-resolved PL (TRPL) spectra were measured with an FS5 spectrofluorometer from Edinburgh Instruments (Livingston, UK). A 478.4 nm laser was used as an excitation source. All measurements were performed in environmental conditions at room temperature.

3. Results and Discussion

3.1. Characterization of Carbon Dots and F8BT/Carbon Dot Interfaces

In this work, we studied carbon and nitrogen-functionalized carbon nanodots that were synthesized by a bottom-up approach from inexpensive, organic precursors; their design was based on sodium carboxylate groups and carboxylic acid with amino-terminated groups, respectively. The surface terminal groups of the synthesized carbon dots (C-dots) and nitrogen-doped carbon dots (NC-dots) are illustrated in Figure 1a. They were both processed from solutions in orthogonal to organic semiconductor solvents, such as methanol and dimethylformamide (DMF).

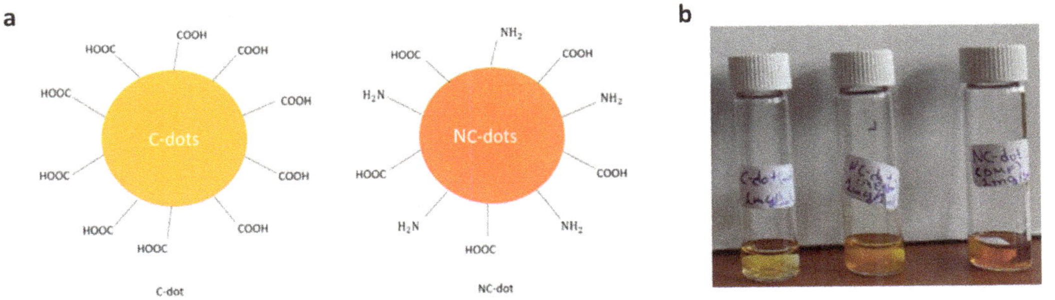

Figure 1. (**a**) Illustration of the chemical structures (surface functionalization) of C-dots and NC-dots. (**b**) Photographs demonstrating the solution processability of carbon nanodots. From left to the right: a methanol solution of C-dots, a methanol solution of NC-dots and a DMF solution of NC-dots.

The morphology and structure of these nanodots and carbon dots were investigated by transmission electron microscopy (TEM). Figure 2a shows a TEM image of the C-dots, suggesting the formation of uniform spherical particles with an average size of around 3 nm. The crystallinity of the nanodots was studied by X-ray diffraction (XRD) measurements. Figure 2b,c show the XRD patterns of thin film deposited from C-dot and NC-dot solutions, respectively. In the case of C-dots, the only peak that appeared was assigned to the silicon substrate; therefore, the sample was amorphous. However, the diffraction pattern of NC-dots consisted of more intense peaks that revealed the formation of a more crystalline structure in these dots. As represented in Figure 2c, the XRD pattern of the NC-dots included peaks at 2θ = 20.9°, 28.7°, 39.0° and 61.6°. The corresponding Miller indices according to the relevant literature are also shown in Figure 2c,d [27]. Notably, XRD revealed that the NC-dot sample was more crystalline compared to the pristine C-dots. The lattice d-spacing was found to be 0.31 nm. This could be beneficial for electron transport when these materials are used as cathode interfacial layers in organic optoelectronic devices. Figure 2d depicts the UV–Vis absorption spectra and the derived Tauc plot for the estimation of the energy bandgap (E_G) values of these materials. The extracted values are 2.6 eV for C-dots and 2.3 eV for the NC-dots.

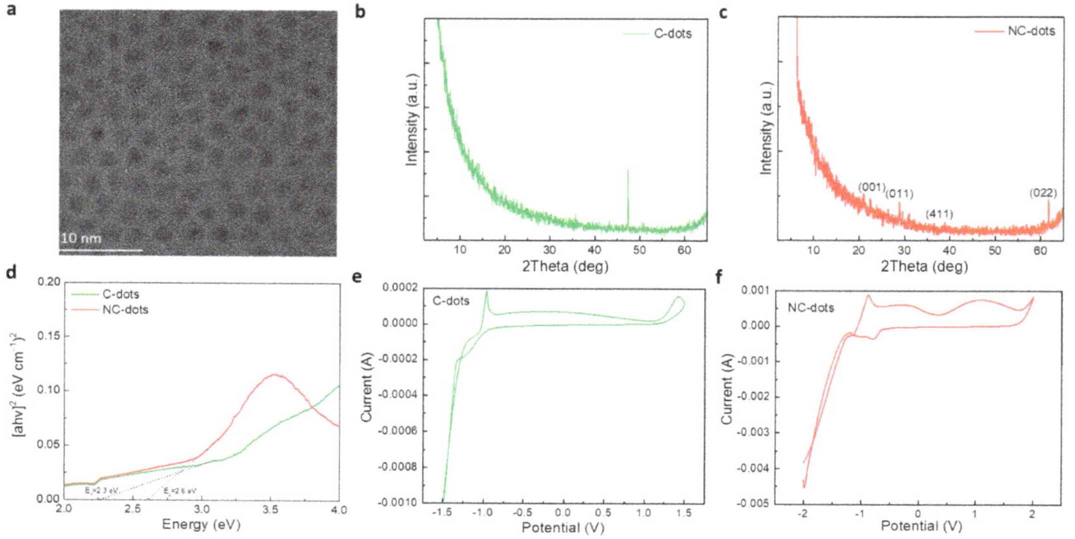

Figure 2. (**a**) Transmission electron microscopy (TEM) images of carbon dots. XRD patterns of (**b**) C-dot and (**c**) NC-dot films drop cast on silicon substrates from concentrated (10 mg mL^{-1}) methanol solutions. (**d**) Tauc plots of C-dot and NC-dot films drop cast onto quartz substrates from concentrated (10 mg mL^{-1}) methanol solutions. Cyclic voltammetry of (**e**) C-dot and (**f**) NC-dot films coated on indium-tin oxide (ITO)/glass substrates at a scan rate of 0.1 V s^{-1} in a 0.1 M $LiClO_4$ aqueous electrolyte solution.

Amino groups have been considered beneficial for use in EILs as they usually induce the formation of negative interfacial dipoles at the cathode interface [28]. As a result, they can reduce the electron injection barrier at the respective contact. Therefore, it is of significant importance to estimate the energy levels of the synthesized materials and obtain valuable information about the energetic alignment at the cathode interface. As cyclic voltammetry is considered one of the most powerful tools to investigate the reduction and oxidation processes of molecular species, we employed this method to estimate the highest occupied molecular orbital (HOMO) and lowest unoccupied molecular orbital (LUMO) energy levels of these materials. Figure 2e,f show the cyclic voltammograms of C-dot and NC-dot films, respectively, deposited on indium-tin oxide (ITO)/glass substrates.

Oxidation and reduction processes are represented, allowing the HOMO and LUMO calculation, respectively, using the empirical formulas [29,30]:

$$E_{HOMO} = -(E_{ox,onset} + 4.4)\ \text{eV} \tag{1}$$

$$E_{LUMO} = -(E_{red,onset} + 4.4)\ \text{eV} \tag{2}$$

where $E_{ox,onset}$ and $E_{red,onset}$ are the oxidation and reduction potential onset, respectively, defined as the position where the current starts to differ from the baseline. From the cyclic voltammograms the extracted $E_{ox,onset}$ value is +1.2 V and +1.8 V for C-dots and NC-dots, respectively, which results in HOMO levels of −5.6 eV for C-dots and −6.2 eV for NC-dots. The LUMO levels of C-dots and NC-dots are −3.4 eV and −3.5eV, respectively, determined by the reduction potential onset (−1 V for C-dots and −0.9 V for NC-dots).

The deposited films were further characterized by Fourier-transform infrared spectroscopy (FT-IR) in order to recognize any possible changes in the F8BT film after the deposition of C-dots and NC-dots (Figure 3). In the C-dot spectrum, the broad band observed in the range of 3500-3100 cm^{-1} corresponds to the stretching vibration of the O–H bond, while the weak band at 2930 cm^{-1} corresponds to C–H stretching. The two bands at 1560 cm^{-1} and 1395 cm^{-1} correspond to the asymmetric and symmetric stretching of the carboxylate group. In the case of the NC-dot spectrum, the stretching modes of N–H, O–H and C–H are detected in the broad band ranging from 3600 cm^{-1} to 2800 cm^{-1}. The other bands detected at 1700 cm^{-1}, 1650 cm^{-1} and 1560 cm^{-1} refer to the C=C stretching mode, the C=O stretching mode and the N-H bending mode, respectively.

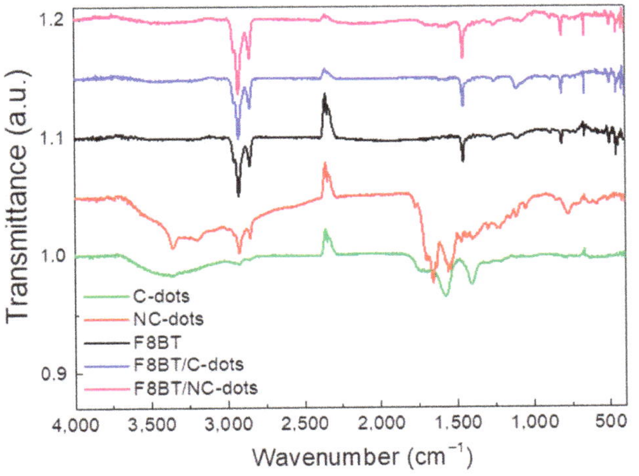

Figure 3. FTIR spectra of pristine carbon dots and F8BT films and of F8BT/C-dot and F8BT/NC-dot interfaces.

As far as the spectrum of the F8BT emitter is concerned, the bands at 2930 cm^{-1} and 2850 cm^{-1} correspond to the stretching mode of the C–H bond, while the band at 1460 cm^{-1} is due to the C=C stretching mode of an aromatic group [31,32]. Furthermore, the stretching mode between the phenyl rings is detected at 1250 cm^{-1}, the deformation of aliphatic chain bonds at 1100 cm^{-1} and the C–H rocking mode at 812 cm^{-1}. In the case of the F8BT films with C-dots deposited on top, a slight increase in the intensity and width of the band at 1100 cm^{-1} is observed, with no other obvious changes. When NC-dots are deposited on top of a F8BT film [32], a similar change is apparent in the 1100 cm^{-1} region, where the band corresponding to aliphatic chain deformation almost disappears. This is a sign of possible interaction between the aliphatic chains of the F8BT molecule and the NC-dot material deposited on top. However, the bands corresponding to the stretching vibrations of the same aliphatic chains do not change, which is an indication of the limited extent of this interaction.

We next investigated the optoelectronic and morphological properties of F8BT before and after coverage with thin carbon dot interlayers (processed from 0.5 mg mL^{-1} in methanol solutions). In Figure 4a, the UV–Vis absorption spectra of F8BT without and with carbon dots spin-coated are presented. All these spectra exhibit absorption peaks centered at wavelengths of 372 nm and 450 nm, corresponding to F8BT [33]. The characteristic carbon dot absorption peaks were not detected in layers deposited on top of the F8BT, which is an indication that they form very thin interlayers on top of F8BT.

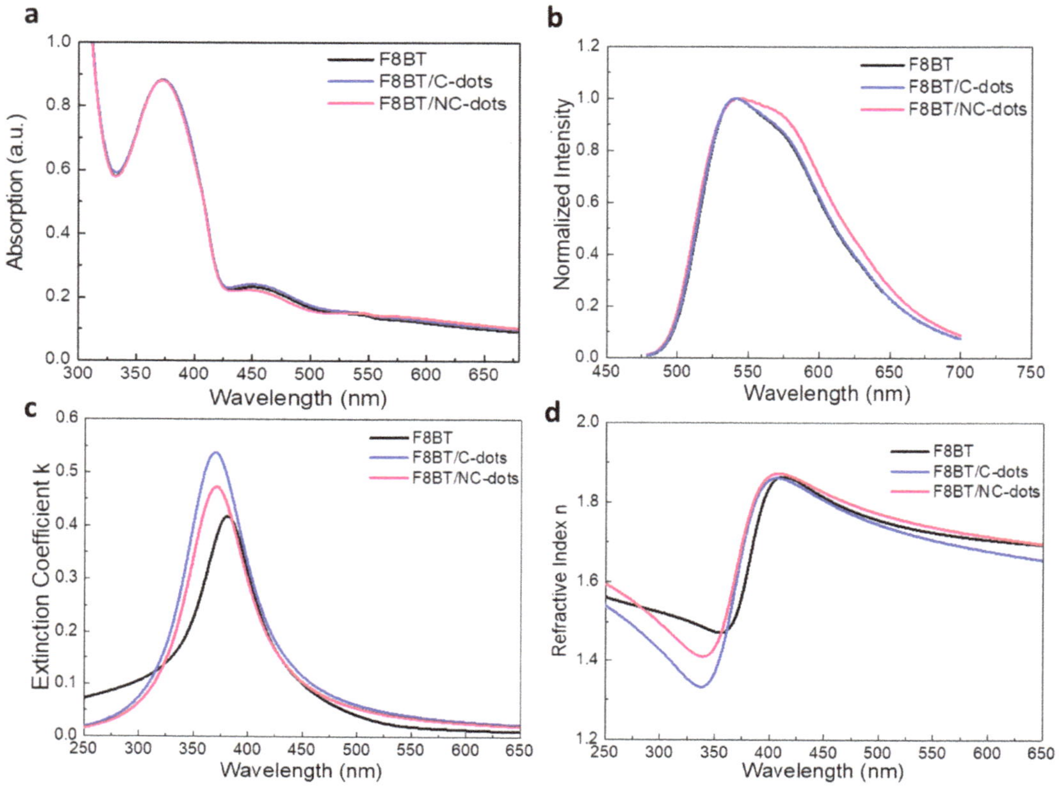

Figure 4. (**a**) UV–Vis absorption and (**b**) steady-state PL spectra of F8BT pristine and coated with C-dots and NC-dots. (**c**) Extinction coefficient and (**d**) refractive index measurements of F8BT and F8BT/C-dot and F8BT/NC-dot interfaces.

Moreover, the emission characteristics of the same samples were investigated by measuring the steady-state photoluminescence (PL) spectra (Figure 4b). While the spectra of F8BT and F8BT/C-dots are identical, a shoulder observed at approximately 580 nm for the F8BT/NC-dot sample can be attributed to an interaction between the aliphatic chains of the F8BT molecule and the NC-dot interlayer, as already discussed in the FT-IR measurements. In addition, no changes in the time-resolved photoluminescence (TRPL) spectra (Figure S1, Supplementary Materials) of pristine and nanodot-coated F8BT films are observed, exhibiting similar average lifetimes in all cases. However, the extinction coefficients and refractive indices (RI) of F8BT pristine and coated with carbon dots present some changes (Figure 4c,d). In particular, a change in the RI may be an indication that the insertion of these dots can change the light outcoupling efficiency of the fabricated OLEDs. Such change, however, is expected to be rather marginal since the difference in RI is very small at the emission wavelength range of F8BT (centered on 530 nm). Notably, the carbon dot interlayers induce a small increase in the surface roughness of F8BT, as indicated by the similarity of the atomic force microscopy (AFM) surface topography of these samples (Figure 5). A small increase in surface roughness or the formation of small nanostructures may be beneficial for the device performance as they enlarge the area, where electron injection towards the emissive layer occurs [34–36].

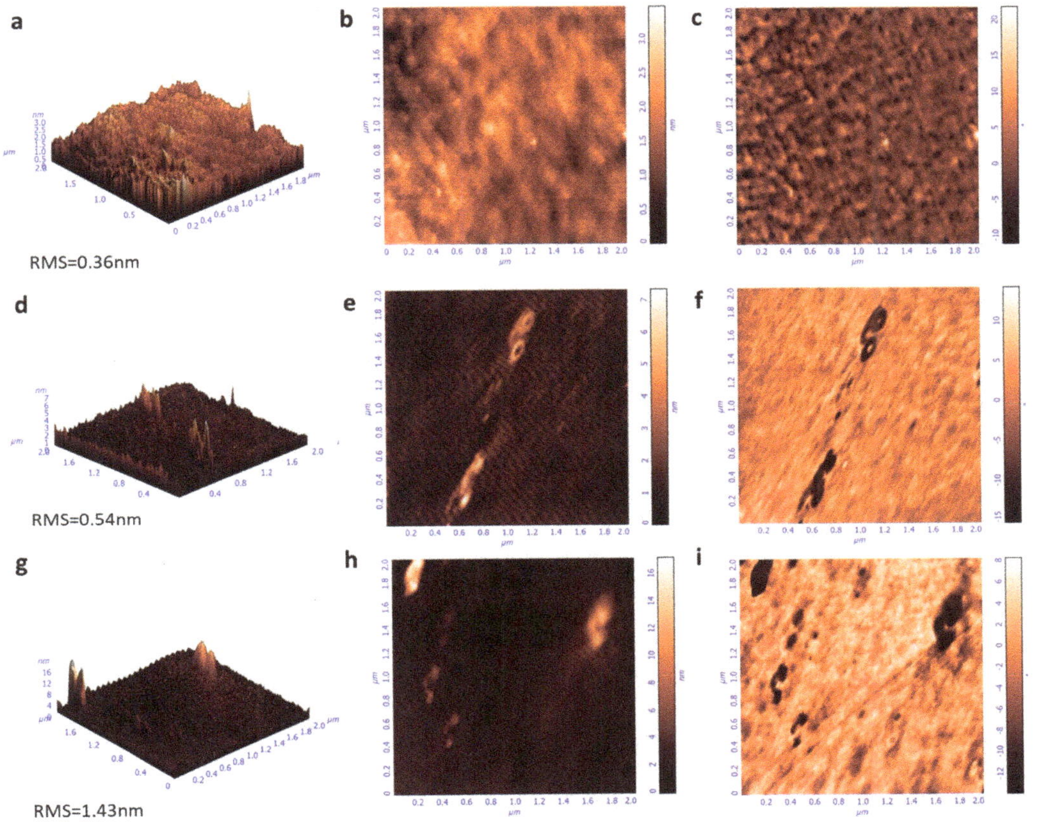

Figure 5. Atomic force microscopy (AFM) surface topographies (height, 3D-left and 2D-middle, phase-right) of (**a–c**) F8BT, (**d–f**) F8BT/C-dot and (**g–i**) F8BT/NC-dot films.

3.2. OLED Measurements

To demonstrate the practical utility of these solution-processed carbon nanodots, we next applied them as EIL/ETLs in fluorescent OLEDs based on a green-yellow fluorescent polymer, well known as F8BT. The accurate structure of the OLEDs is represented in Figure 6a. Specifically, the devices consisted of indium-tin oxide (ITO) coated glass substrates as the transparent anode electrode, poly(3,4-ethylenedioxythiophene) polystyrene sulfonate (PEDOT:PSS) as the hole transport layer, F8BT as the emissive layer, C-dots or NC-dots as the electron transport layers and Al contact as the cathode electrode. A reference device using a commonly used solution-processed EIL, namely cesium carbonate (Cs_2CO_3), was also fabricated. The corresponding energy level diagram of the materials embedded into the OLED structure, as well as the chemical structure of F8BT, is illustrated in Figure 6b. Note that the work functions of ITO, PEDOT:PSS and Al, along with the energy levels of F8BT, were taken from the literature [37,38]. The HOMO and LUMO levels of both carbon nanodots were estimated by cyclic voltammetry and optical measurements, as already mentioned.

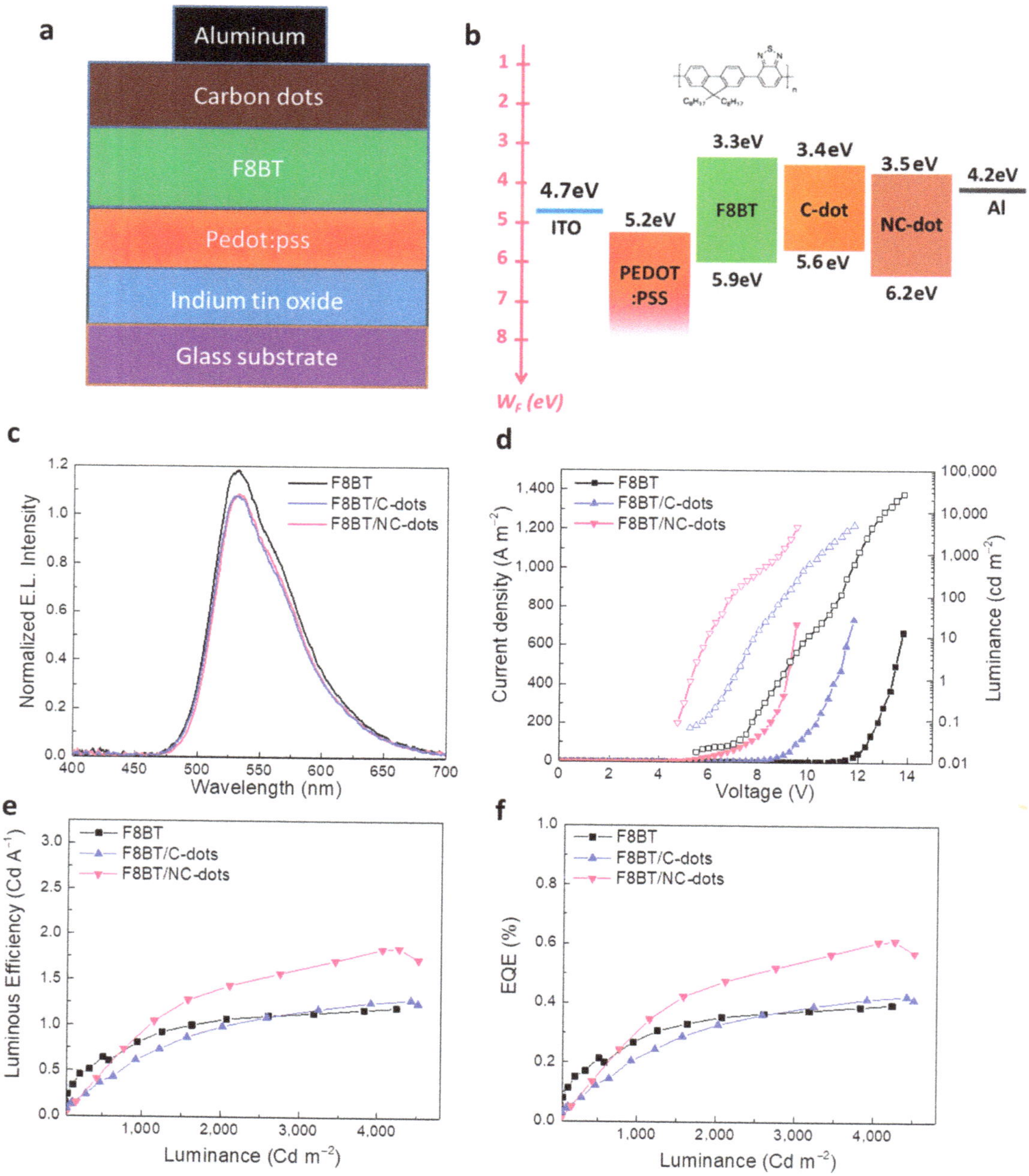

Figure 6. (**a**) The OLED architecture and (**b**) the corresponding energy level diagram. The chemical structure of the F8BT is also presented. (**c**) Electroluminescence (EL) spectra of the different devices at a voltage of 5 V. (**d**) Current density (solid symbols)–voltage and luminance (open symbols)–voltage characteristics of the three types of OLEDs. (**e**) Luminous efficiency (LE) and (**f**) external quantum efficiency (EQE) of the same devices.

Figure 6c shows the electroluminescence (EL) spectra of the three different types of devices. It becomes obvious that the devices with C-dots and NC-dots do not affect the spectral dependence of the emission, which originates only from F8BT. Figure 6d illustrates a plot of the current density versus the applied voltage and luminance versus the applied voltage of the OLEDs. Devices with C-dots and, especially, those using NC-dots have a rapid increase in the current density at the low-voltage regime—an indication of the reduced electron injection barrier in these devices compared to the reference one. Furthermore, from the plot of luminance versus the applied voltage of these OLEDs, the device turn-on voltage (V_{ON}, the voltage where the luminance becomes equal to 1 cd m^{-2}) can be estimated. For the reference device, the V_{ON} is calculated to be around 8 V, whereas it is decreased to nearly 6 V and further to 4 V for the devices using C-dots and NC-dots, respectively, indicating the crucial role of the carbon dots in the device performance. Figure 6e,f present the luminous efficiency (LE) and external quantum efficiency (EQE) of the prepared OLEDs, where a clear enhancement in the device performance is observed when nanodots, especially NC-dots, are incorporated in the device. In particular, the NC-dot-based OLED exhibits a ~1.5 fold higher LE and EQE compared with the reference device. A small enhancement in hydrophobicity induced by the coverage of F8BT with the carbon nanodots (Figure S2, Supplementary Materials) is also observed, which could be beneficial in the device's stability.

Furthermore, to unambiguously demonstrate the beneficial effect of these carbon dots for OLEDs, mainly due to a reduction in the electron injection barrier, we fabricated electron-only devices and measured the carrier obtained by blocking one type of carrier (holes) through the absence of the hole injection layer (PEDOT:PSS) at the anode side of the device. Figure 7 shows the current density taken in electron-only devices using either Cs_2CO_3 or the newly developed carbon nanodots at one side of the device. By considering that the bulk-limited currents are similar to these devices, the obtained current densities are governed by electrode-limited processes (electron barriers are responsible for conduction) [16,39]. As shown in Figure 7, the current density versus voltage (J-V) curve of the reference diode (shown in a semilog plot) is always lower compared to the current densities of the carbon dots using diodes, a clear indication that the electron injection barrier is reduced upon the application of carbon nanodots, especially those with the amino-functionalized surface, as cathode interlayers in these devices.

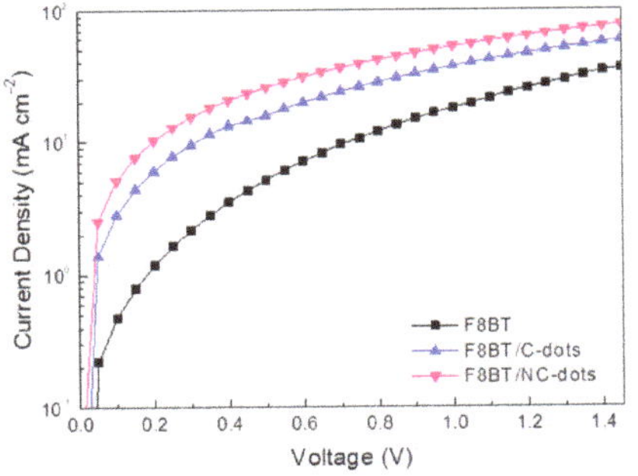

Figure 7. Electron-only current densities of diodes with the structure ITO/F8BT/Al without and with C-dots and NC-dots at the cathode side of the device.

3.3. OSC Measurements

In order to strengthen our approach, we also employed the nanodots as cathode interlayers in inverted OSCs based on a PTB7-Th (poly[4,8-bis(5-(2-ethylhexyl)thiophen-2-yl)benzo[1,2-b;4,5-b′]dithiophene-2,6-diyl-alt-(4-(2-ethylhexyl)-3-fluorothieno[3,4-b]thiophene-)-2-carboxylate-2-6-diyl)]) donor and PCBM ([6,6]-phenyl-C61-butyric acid methyl ester) acceptor blended photoactive layer. Figure 8a illustrates the device structure of the fabricated inverted OSCs, while Figure 8b presents the chemical structure of the donor and acceptor materials that constituted the photoactive layer. Carbon nanodots spin-coated from a methanol solution with a concentration of 0.5 mg mL^{-1} on tin oxide (SnO_2) were used as ETLs, forming an ultra-thin film to improve the electron transport between the cathode and the photoactive layer. Note that molybdenum oxide (MoO_x), used as a hole transport layer, and a reference device without the nanodots were also considered for comparison. Figure 8c shows the transmittance spectra of the pristine and nanodot-coated SnO_2. It is observed that the nanodot interlayers slightly affect the transmittance and absorption (Figure S3a) properties of SnO_2 in the visible region, suggesting no prevention of sunlight reaching the photoactive layer, which could lead to a current density reduction. Accordingly, no changes in energy bandgap values (E_G = 3.65 eV) are observed for SnO_2 films coated with or without carbon dots (Figure S3b).

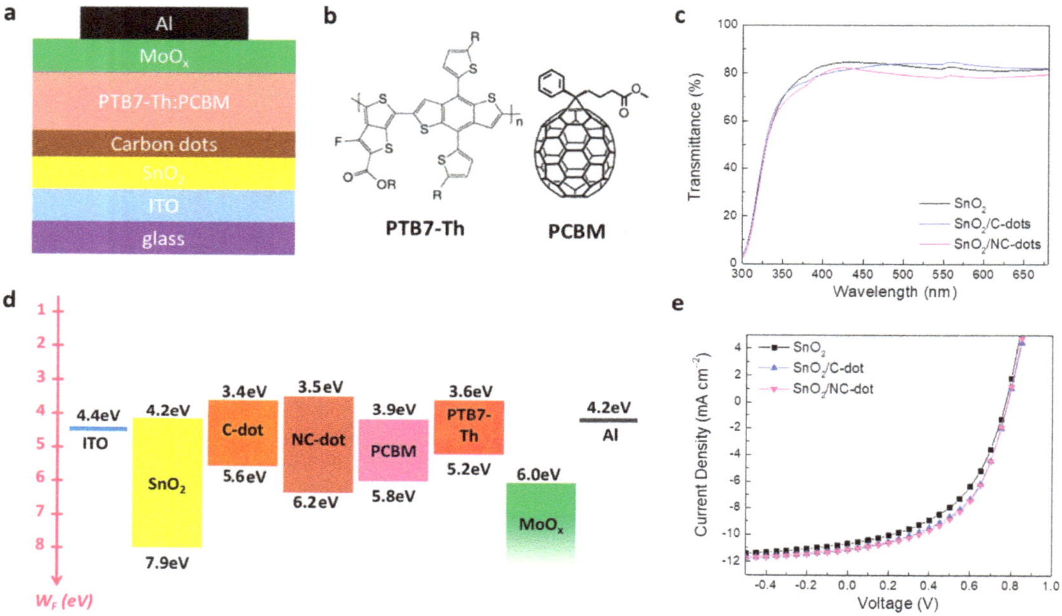

Figure 8. (**a**) Schematic representation of the fabricated inverted OSCs. (**b**) Chemical structure of the donor and acceptor materials used in the photoactive blended layer. (**c**) Transmittance spectra of the SnO_2 film coated with or without the C-dots and NC-dots. (**d**) Energy level diagram of the different layers used for the fabrication of the inverted OSCs. (**e**) Current density–voltage (J-V) characteristic curves measured under AM 1.5G illumination.

Figure 8d represents the energy level diagram of the materials used in the inverted OSCs. Note that, except the energy levels of the carbon nanodots, the work functions and energy levels of the SnO_2, PTB7-Th and PCBM were taken from previous work [40,41]. As in the case of OLEDs, a reduction in the electron extraction barrier at the cathode/PCBM interface is observed, verifying the beneficial role of the nanodots coated atop the SnO_2 film as cathode interlayers in the device performance. The current density–voltage (J-V) characteristic curves of the nanodot-based and reference OSCs are depicted in Figure 8e. A ~13% improvement in the device efficiency is observed for the OSCs with the nanodot interlayers, revealing the universality of the carbon dots used as ETLs in organic opto-electronic devices. In particular, the NC-dot-based device exhibited a short-circuit current density (J_{SC}) of 11.14 mA cm^{-2}, open-circuit voltage (V_{OC}) of 0.78 V and fill factor (FF) of 0.52, resulting in a higher PCE of 4.52% compared with the reference cell, showing a PCE

of 3.94% (J_{SC} of 10.66 mA cm^{-2}, V_{OC} of 0.77 V and FF of 0.48). The device with the C-dots showed also higher electrical parameters (J_{SC} of 11.08 mA cm^{-2}, V_{OC} of 0.78 V and FF of 0.51) and thus efficiency (PCE of 4.41%) with respect to the reference device.

4. Conclusions

We have demonstrated here the beneficial role of carbon nanodots with different surface functionalization in the performance of solution-processed green OLEDs and inverted OSCs. These materials were investigated for their optoelectronic properties and possible interaction with the organic semiconductors. Our experimental data indicate that they induce significant reductions in the electron injection/extraction barrier, therefore resulting in the fabrication of OLEDs and OSCs with significantly reduced turn-on voltages and increased short-circuit current density, respectively. Our results demonstrate the beneficial effects of abundant, carbon-based materials in organic optoelectronics for large-area applications.

Supplementary Materials: The following supporting information can be downloaded at: https://www.mdpi.com/article/10.3390/nano13010169/s1, Figure S1: Time-resolved photoluminescence (TRPL) spectra of F8BT, F8BT/C-dots, and F8BT/NC-dots; Figure S2: Water contact angle measurements taken in (a) F8BT, (b) F8BT/C-dots and (c) F8BT/NC-dots; Figure S3: (a) Absorption spectra and (b) tauc plots of pristine and nanodots-coated SnO2.

Author Contributions: Investigation, Z.G., A.V., K.L., K.M., G.C., K.-K.A. and E.K.E.; Writing-Original Draft, Z.G.; Methodology, L.C.P., E.K.E., S.G., A.R.b.M.Y. and K.A.; Formal Analysis, A.C.; Data Curation, A.C.; Conceptualization, A.G.C.; Writing-Review and editing, A.G.C., K.A. and A.S.; Supervision, M.V.; Project Administration, M.V.; Visualization, A.S. All authors have read and agreed to the published version of the manuscript.

Funding: This research was funded by the European Regional Development Fund of the European Union and Greek national funds through the Operational Program Competitiveness, Entrepreneurship and Innovation, under the call RESEARCH–CREATE–INNOVATE (project code:T2-EDK-04175-OLED-LUMIN-FOOD).

Data Availability Statement: The data that support the findings of this study can become available by the corresponding authors upon reasonable request.

Acknowledgments: This research has been co-financed by the European Regional Development Fund of the European Union and Greek national funds through the Operational Program Competitiveness, Entrepreneurship and Innovation, under the call RESEARCH–CREATE–INNOVATE (project code: T2-EDK-04175-OLED-LUMIN-FOOD).

Conflicts of Interest: The authors declare no conflict of interest.

References

1. Tang, C.W.; VanSlyke, S.A. Organic Electroluminescent Diodes. *Appl. Phys. Lett.* **1987**, *51*, 913–915. [CrossRef]
2. Steckl, A.J.; Heikenfeld, J.; Allen, S.C. Light Wave Coupled Flat Panel Displays and Solid-State Lighting Using Hybrid Inorganic/Organic Materials. *J. Disp. Technol.* **2005**, *1*, 157–166. [CrossRef]
3. Davids, P.S.; Kogan, S.M.; Parker, I.D.; Smith, D.L. Charge Injection in Organic Light-emitting Diodes: Tunneling into Low Mobility Materials. *Appl. Phys. Lett.* **1996**, *69*, 2270–2272. [CrossRef]
4. Ishii, H.; Sugiyama, K.; Ito, E.; Seki, K. Energy Level Alignment and Interfacial Electronic Structures at Organic/Metal and Organic/Organic Interfaces. *Adv. Mater.* **1999**, *11*, 605–625. [CrossRef]
5. Yang, S.-M.; Jang, S.G.; Choi, D.-G.; Kim, S.; Yu, H.K. Nanomachining by Colloidal Lithography. *Small* **2006**, *2*, 458–475. [CrossRef] [PubMed]
6. Khan, M.A.; Xu, W.; Wei, F.; Bai, Y.; Jiang, X.Y.; Zhang, Z.L.; Zhu, W.Q. Highly Efficient Organic Electroluminescent Diodes Realized by Efficient Charge Balance with Optimized Electron and Hole Transport Layers. *Solid State Commun.* **2007**, *144*, 343–346. [CrossRef]
7. Patel, N.K.; Cina, S.; Burroughes, J.H. High-Efficiency Organic Light-Emitting Diodes. *IEEE J. Sel. Top. Quantum Electron.* **2002**, *8*, 346–361. [CrossRef]
8. Qi, B.; Zhang, Z.G.; Wang, J. Uncovering the Role of Cathode Buffer Layer in Organic Solar Cells. *Sci. Rep.* **2015**, *5*, 7803. [CrossRef]

9. Kadam, K.D.; Kim, H.; Rehman, S.; Patil, H.; Aziz, J.; Dongale, T.D.; Khan, M.F.; Kim, D.-K. Optimization of ZnO:PEIE as an Electron Transport Layer for Flexible Organic Solar Cells. *Energy Fuels* **2021**, *35*, 12416–12424. [CrossRef]
10. Liu, C.; Xiao, C.; Li, W. Zinc Oxide Nanoparticles as Electron Transporting Interlayer inOrganic Solar Cells. *J. Mater. Chem. C* **2021**, *9*, 14093–14114. [CrossRef]
11. Özgür, Ü.; Alivov, Y.I.; Liu, C.; Teke, A.; Reshchikov, M.A.; Doğan, S.; Avrutin, V.; Cho, S.-J.; Morkoç, H. A Comprehensive Review of ZnO Materials and Devices. *J. Appl. Phys.* **2005**, *98*, 041301. [CrossRef]
12. Lee, H.; Park, I.; Kwak, J.; Yoon, D.Y.; Lee, C. Improvement of Electron Injection in Inverted Bottom-Emission Blue Phosphorescent Organic Light Emitting Diodes Using Zinc Oxide Nanoparticles. *Appl. Phys. Lett.* **2010**, *96*, 153306. [CrossRef]
13. Huang, S.; Ali, N.; Huai, Z.; Ren, J.; Sun, Y.; Zhao, X.; Fu, G.; Kong, W.; Yang, S. A Facile Strategy for Enhanced Performance of Inverted Organic Solar Cells Based on Low-Temperature Solution-Processed SnO_2 Electron Transport Layer. *Org. Electron.* **2020**, *78*, 105555. [CrossRef]
14. Lee, H.; Kang, C.-M.; Park, M.; Kwak, J.; Lee, C. Improved Efficiency of Inverted Light-Emitting Diodes Using Tin Dioxide Nanoparticles as an Electron Injection Layer. *ACS Appl. Mater. Interfaces* **2013**, *5*, 1977–1981. [CrossRef]
15. Li, Q.; Liu, H.; Sun, W.; Wang, S.; Dong, X.; Wang, L.; Li, X. Hole-Transporting Material Based on Spirobifluorene Unit with Perfect Amorphous and High Stability for Efficient OLEDs. *J. Mater. Sci. Mater. Electron.* **2019**, *30*, 11440–11450. [CrossRef]
16. Ma, D.; Chen, Y. *Organic Semiconductor Heterojunctions and Its Application in Organic Light-Emitting Diodes*; Springer Series in Materials Science; Springer: Berlin/Heidelberg, Germany, 2017; Volume 250, ISBN 978-3-662-53693-3.
17. Verykios, A.; Papadakis, M.; Soultati, A.; Skoulikidou, M.-C.; Papaioannou, G.; Gardelis, S.; Petsalakis, I.D.; Theodorakopoulos, G.; Petropoulos, V.; Palilis, L.C.; et al. Functionalized Zinc Porphyrins with Various Peripheral Groups for Interfacial Electron Injection Barrier Control in Organic Light Emitting Diodes. *ACS Omega* **2018**, *3*, 10008–10018. [CrossRef]
18. Kalinowski, J. *Organic Light-Emitting Diodes: Principles, Characteristics, and Processes*; Optical Engineering; Marcel Dekker: New York, NY, USA, 2005; ISBN 978-0-8247-5947-6.
19. Soultati, A.; Papadimitropoulos, G.; Davazoglou, D.; Argitis, P.; Alexandropoulos, D.; Politi, C.T.; Vainos, N.; Pistolis, G.; Coutsolelos, A.G.; Vasilopoulou, M. Near-IR Organic Light Emitting Diodes Based on Porphyrin Compounds. In Proceedings of the 2015 17th International Conference on Transparent Optical Networks (ICTON), Budapest, Hungary, 5–9 July 2015; pp. 1–4.
20. Kido, J.; Ohtaki, C.; Hongawa, K.; Okuyama, K.; Nagai, K. 1,2,4-Triazole Derivative as an Electron Transport Layer in Organic Electroluminescent Devices. *Jpn. J. Appl. Phys.* **1993**, *32*, L917–L920. [CrossRef]
21. Kido, J.; Hongawa, K.; Okuyama, K.; Nagai, K. Bright Blue Electroluminescence from Poly(*N*-vinylcarbazole). *Appl. Phys. Lett.* **1993**, *63*, 2627–2629. [CrossRef]
22. Yu, G.; Yin, S.; Liu, Y.; Shuai, Z.; Zhu, D. Structures, Electronic States, and Electroluminescent Properties of a Zinc(II) 2-(2-Hydroxyphenyl)Benzothiazolate Complex. *J. Am. Chem. Soc.* **2003**, *125*, 14816–14824. [CrossRef]
23. Kido, J.; Kimura, M.; Nagai, K. Multilayer White Light-Emitting Organic Electroluminescent Device. *Science* **1995**, *267*, 1332–1334. [CrossRef]
24. Dailey, S.; Halim, M.; Rebourt, E.; Horsburgh, L.E.; Samuel, I.D.W.; Monkman, A.P. An Efficient Electron-Transporting Polymer for Light-Emitting Diodes. *J. Phys. Condens. Matter* **1998**, *10*, 5171–5178. [CrossRef]
25. Yamamoto, T.; Maruyama, T.; Zhou, Z.-H.; Ito, T.; Fukuda, T.; Yoneda, Y.; Begum, F.; Ikeda, T.; Sasaki, S. Pi.-Conjugated Poly(Pyridine-2,5-Diyl), Poly(2,2′-Bipyridine-5,5′-Diyl), and Their Alkyl Derivatives. Preparation, Linear Structure, Function as a Ligand to Form Their Transition Metal Complexes, Catalytic Reactions, n-Type Electrically Conducting Properties, Optical Properties, and Alignment on Substrates. *J. Am. Chem. Soc.* **1994**, *116*, 4832–4845. [CrossRef]
26. Vasilopoulou, M.; Douvas, A.M.; Georgiadou, D.G.; Palilis, L.C.; Kennou, S.; Sygellou, L.; Soultati, A.; Kostis, I.; Papadimitropoulos, G.; Argitis, P. The Influence of Hydrogeneration and Oxygen Vacancies on Molybdenum Oxides Work Function and Gap States for Application in Organic Optoelectronics. *J. Am. Chem. Soc.* **2012**, *134*, 16178–16187. [CrossRef] [PubMed]
27. Pal, A.; Natu, G.; Ahmad, K.; Chattopadhyaya, A. Phosphorus Induced Crystallinity in Carbon Dots for Solar Light Assisted Seawater Desalination. *J. Mater. Chem. A* **2018**, *6*, 4111–4118. [CrossRef]
28. Lee, B.R.; Lee, S.; Park, J.H.; Jung, E.D.; Yu, J.C.; Nam, Y.S.; Heo, J.; Kim, J.-Y.; Kim, B.-S.; Song, M.H. Amine-Based Interfacial Molecules for Inverted Polymer-Based Optoelectronic Devices. *Adv. Mater.* **2015**, *27*, 3553–3559. [CrossRef]
29. Omer, K.M.; Tofiq, D.I.; Hassan, A.Q. Solvothermal Synthesis of Phosphorus and Nitrogen Doped Carbon Quantum Dots as a Fluorescent Probe for Iron(III). *Microchim. Acta* **2018**, *185*, 466. [CrossRef]
30. Zhang, H.; Chen, Y.; Liang, M.; Xu, L.; Qi, S.; Chen, H.; Chen, X. Solid-Phase Synthesis of Highly Fluorescent Nitrogen-Doped Carbon Dots for Sensitive and Selective Probing Ferric Ions in Living Cells. *Anal. Chem.* **2014**, *86*, 9846–9852. [CrossRef]
31. de Amorim Lima, N.; Camargo Avila, H.; do Nascimento Marchiori, C.; Gondim Sampaio, S.; Ferreira Mota, J.; Gomes Pereira Ribeiro, V.; da Silva Clemente, C.; Mele, G.; Cremona, M.; Mazzetto, S. Light-Emitting Porphyrin Derivative Obtained from a Subproduct of the Cashew Nut Shell Liquid: A Promising Material for OLED Applications. *Materials* **2019**, *12*, 1063. [CrossRef]
32. Sultan Erkan Theoretical and Experimental Spectroscopic Properties and Molecular Docking of F8BT P-Type Semiconducting Polymer. *Russ. J. Phys. Chem. A* **2020**, *94*, 445–452. [CrossRef]
33. Ayesta, I.; Illarramendi, M.A.; Arrue, J.; Jimenez, F.; Zubia, J.; Bikandi, I.; Ugartemendia, J.M.; Sarasua, J.-R. Luminescence Study of Polymer Optical Fibers Doped With Conjugated Polymers. *J. Light. Technol.* **2012**, *30*, 3367–3375. [CrossRef]
34. Jeong, K.W.; Kim, H.S.; Yi, G.R.; Kim, C.K. Enhancing the Electroluminescence of OLEDs by Using ZnO Nanoparticle Electron Transport Layers That Exhibit the Auger Electron Effect. *Mol. Cryst. Liq. Cryst.* **2018**, *663*, 61–70. [CrossRef]

35. Li, W.; Wu, X.; Liu, G.; Li, Y.; Wu, L.; Fu, B.; Wang, W.; Zhang, D.; Zhao, J. Enhanced Electron Transportation of PF-NR2 Cathode Interface by Gold Nanoparticles. *Nanoscale Res. Lett.* **2019**, *14*, 261. [CrossRef] [PubMed]
36. Sun, H.; Chen, Y.; Zhu, L.; Guo, Q.; Yang, D.; Chen, J.; Ma, D. Realization of Optimal Interconnector for Tandem Organic Light-Emitting Diodes with Record Efficiency. *Adv. Electron. Mater.* **2015**, *1*, 1500176. [CrossRef]
37. Sardar, G.; Shukla, A.; Moore, E.G.; Banappanavar, G.; Lo, S.-C.; Namdas, E.B.; Kabra, D. Reduced Singlet-Triplet Annihilation for Low Threshold Aplified Spontaneous Emission from a Blue Polyfluorene Electroluminescent Organic Semiconductor. *J. Phys. Chem. C* **2022**, *126*, 9069–9075. [CrossRef]
38. Morii, K.; Fukagawa, H. Hybrid Organic-Inorganic Light-Emitting Diode. In *Air-Stable Inverted Organic Light-Emitting Diodes*; SpringerBriefs in Applied Sciences and Technology; Springer: Cham, Switzerland, 2020. [CrossRef]
39. Giebink, N.C.; Wiederrecht, G.P.; Wasielewski, M.R.; Forrest, S.R. Ideal Diode Equation for Organic Heterojunctions. I. Derivation and Application. *Phys. Rev. B* **2010**, *82*, 155305. [CrossRef]
40. Soultati, A.; Verykios, A.; Panagiotakis, S.; Armadorou, K.-K.; Haider, M.I.; Kaltzoglou, A.; Drivas, C.; Fakharuddin, A.; Bao, X.; Yang, C.; et al. Suppressing the Photocatalytic Activity of Zinc Oxide Electron-Transport Layer in Nonfullerene Organic Solar Cells with a Pyrene-Bodipy Interlayer. *ACS Appl. Mater. Interfaces* **2020**, *12*, 21961–21973. [CrossRef] [PubMed]
41. Guo, F.; Sun, X.; Ynag, Z.; Wei, J.; Xu, D. Enhanced Lifetime and Photostability with Low-Temperature Mesoporous $ZnTiO_3$/Compact SnO_2 Electrodes in Perovskite Solar Cells. *Angew. Chem.* **2019**, *58*, 18460–18465. [CrossRef]

Disclaimer/Publisher's Note: The statements, opinions and data contained in all publications are solely those of the individual author(s) and contributor(s) and not of MDPI and/or the editor(s). MDPI and/or the editor(s) disclaim responsibility for any injury to people or property resulting from any ideas, methods, instructions or products referred to in the content.

nanomaterials

Article

Modulation of Casimir Force between Graphene-Covered Hyperbolic Materials

Ge Song, Zhixiang Liu *, Lingchun Jia *, Cong Li and Yingli Chang

College of Information Technology, Shanghai Ocean University, Shanghai 201306, China;
gsong@shou.edu.cn (G.S.); cli@shou.edu.cn (C.L.); ylchang@shou.edu.cn (Y.C.)
* Correspondence: zxliu@shou.edu.cn (Z.L.); lcjia@shou.edu.cn (L.J.)

Abstract: A flexible method for modulating the Casimir force is proposed by combining graphene and hyperbolic materials (HMs). The proposed structure employs two candidates other than graphene. One is hexagonal boron nitride (hBN), a natural HM. The other is porous silicon carbide (SiC), which can be treated as an artificial HM by the effective medium theory. The Casimir force between graphene-covered hBN (porous SiC) bulks is presented at zero temperature. The results show that covering HM with graphene increases the Casimir force monotonically. Furthermore, the force can be modulated by varying the Fermi level, especially at large separation distances. The reflection coefficients are thoroughly investigated, and the enhancement is attributed to the interaction of surface plasmons (SPs) supported by graphene and hyperbolic phonon polaritons (HPhPs) supported by HMs. Moreover, the Casimir force can be controlled by the filling factor of porous SiC. The Casimir force can thus be modulated flexibly by designing desired artificial HMs and tuning the Fermi level. The proposed models have promising applications in practical detection and technological fields.

Keywords: Casimir force; graphene; hyperbolic material

1. Introduction

The Casimir force is an intriguing macroscopic effect caused by the quantum fluctuations of electromagnetic fields. Casimir predicted the existence of attractive forces between two parallel perfectly conducting plates in 1948 [1]. Lifshitz then generalized a theory of forces between two semi-infinite dielectric parallel plates with dispersive and absorptive properties at any temperature [2]. The Casimir effect is still a hot topic with the development of microelectromechanical and nanoelectromechanical systems (MEMS and NEMS). Over the last two decades, special emphasis has been placed on theoretical understanding [3–6] and precise experimental measurements [7–13] of the Casimir effect. In the study of Casimir force, the primary geometric configuration is two parallel plates of natural materials separated by a vacuum gap. In general, the force is too weak for practical detection, so enhancing weak Casimir forces is critical. Furthermore, the force is usually attractive and dominates in the submicrometer regime, where irreversible adhesion of neighboring elements in MEMS and NEMS can occur [14–16]. As a result, modulating the Casimir force is both fundamental and technological [5,17]. The realization of repulsive force is related to the symmetry of electric and magnetic properties of the boundary materials [18,19]. Consequently, using special materials with controllable electromagnetic properties to modulate the Casimir force becomes an interesting topic [4]. Metamaterials, for example, as a type of artificial materials, have unusual electromagnetic properties that natural materials do not have, and are used in cloaking [20], vacuum induced transparency [21], and controlling the Casimir effect [22–24]. Furthermore, saturated ferrite materials [25,26] and topological insulators [27–29] are proposed to modulate the Casimir effect.

Graphene, a two-dimensional sheet of carbon atoms arranged in a hexagonal lattice, has piqued the curiosity of many scientists [30]. The linear dispersion relation near the Dirac point causes an extraordinary response to light [31]. In particular, graphene can support surface plasmons (SPs) in the terahertz to infrared frequency ranges [32]. There has also been extensive research into using graphene to modulate the Casimir effect [32–40]. The plasmonic response of graphene is well understood to be highly dependent on the deposited substrate [41]. As a result, Goos-Hänchen shift [42,43], quantum interference [44], and Casimir friction [32,33] have been investigated in graphene-based models. For example, using graphene-covered hyperbolic materials (HMs) can significantly increase Casimir friction due to the coupling of SPs with hyperbolic phonon polaritons (HPhPs) supported by HMs [33]. This enhancement is active, because graphene's optical conductivity is adjustable and can be controlled by an external field or gate voltage.

Hyperbolic materials have gotten a lot of attention in the last decade because of their unique electromagnetic properties [45]. Diagonal elements of a uniaxial HM's permittivity tensor have opposite signs, resulting in a hyperbolic isofrequency contour for TM polarization [46]. It is possible to achieve ultrahigh propagating wave vectors and surface wave excitation by using HMs. Hexagonal boron nitride (hBN) is a natural HM with hyperbolic responses in the infrared frequency range [47]. Hyperbolic phonon polaritons can be supported by hBN, which has been thoroughly investigated [48], and as microfabrication technology develops, artificial HMs with hyperbolic responses in specific frequency bands can be constructed. In general, alternative metal-dielectric layered structures [49] or a lattice of nanowires embedded in a dielectric matrix [46] can be used to realize artificial HMs. Ingredient materials and their proportions can be used to control the desired electromagnetic properties. As a result, when compared to natural HMs, artificial HMs provide additional methods for modulating HPhPs.

It is well known that the electromagnetic properties of the boundary material can modify the Casimir force [6]. In this paper, we investigate the Casimir force between graphene-covered HMs. The remainder of this paper is organized as follows. Section 2 introduces the two models under consideration here, as well as the Casimir force between two graphene-covered HMs. Section 3 demonstrates the modulation of Casimir force caused by the interaction of SPs and HPhPs. The results show that the Casimir force can be actively modulated by the Fermi level and artificial HMs. Section 4 is where we draw our conclusions.

2. Materials and Methods

Figure 1 depicts the scheme that takes into account two different models. The Casimir force between two identical samples separated by d is investigated in each model. One model's samples are graphene-covered hBN bulks, as shown in Figure 1a, while the other model's samples are graphene-covered porous silicon carbide (SiC) bulks as shown in Figure 1b. The models are in free space, with the $x - y$ plane parallel to the graphene plane. The optical properties of graphene in the low-frequency range and high doping limit are determined by its in-plane conductivity σ, which can be expressed as [42]

$$\sigma(\omega) = \frac{ie^2 E_F}{\pi\hbar^2(\omega + i\tau^{-1})}, \tag{1}$$

under the random phase approximation (RPA). Here, e is the electron charge, E_F is the Fermi level, and $\tau = \mu E_F / e v_F^2$ is the relaxation time caused by electron doping, electron defect, and phonon scattering. The mobility of the graphene charge carriers is $\mu = 10^4 - 10^6 \mathrm{cm^2\,V^{-1}s^{-1}}$, and the Fermi velocity is $v_F = 10^6$ m/s.

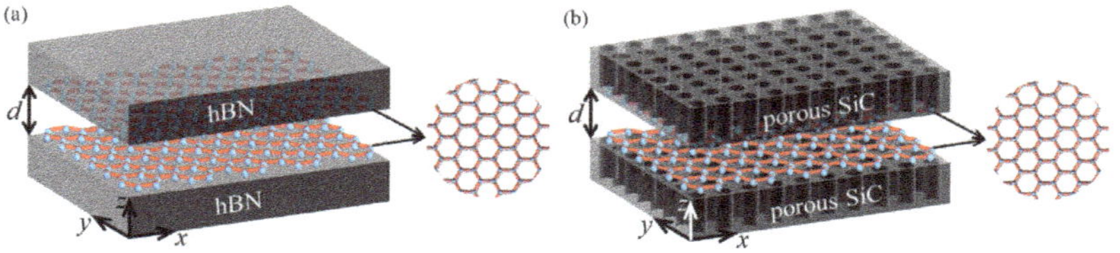

Figure 1. The Casimir force between two graphene-covered HMs separated by *d* is depicted schematically. Two models under consideration: (**a**) graphene-covered hBN and (**b**) graphene-covered porous SiC.

The hBN is a naturally anisotropic material that exhibits hyperbolic dispersion. The permittivity of hBN is a tensor, and the elements of the anisotropic permittivity tensor are as follows [33]

$$\varepsilon_{l,\text{hBN}} = \varepsilon_{l,\infty}\left[1 + \frac{\omega_{\text{LO},l}^2 - \omega_{\text{TO},l}^2}{\omega_{\text{TO},l}^2 - \omega^2 - i\omega\gamma_l}\right], \tag{2}$$

where $l = xx, zz$, LO and TO are two phonon modes, and γ is the damping coefficient. The parameters are $\varepsilon_{xx,\infty} = 4.87$, $\omega_{\text{LO},xx} = 3.0348\ \omega_0$, $\omega_{\text{TO},xx} = 2.5824\ \omega_0$, $\gamma_{xx} = 0.0094\ \omega_0$, $\varepsilon_{zz,\infty} = 2.95$, $\omega_{\text{LO},zz} = 1.5645\ \omega_0$, $\omega_{\text{TO},zz} = 1.4703\ \omega_0$, $\gamma_{zz} = 0.0075\ \omega_0$, and $\omega_0 = 10^{14}$ rad/s. The real parts of $\varepsilon_{l,\text{hBN}}$ are plotted as functions of ω in Figure 2a. Two grey shadow zones satisfying $\text{Re}(\varepsilon_{xx})\text{Re}(\varepsilon_{zz}) < 0$ are obtained, in which HPhPs can be excited. Electromagnetic waves possess a high wave vector in such bands, thus the large electromagnetic local density of the state can be obtained [46].

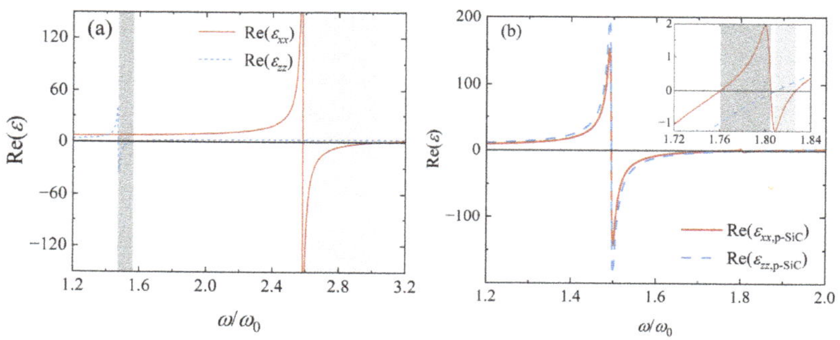

Figure 2. (**a**) The relationship between the real parts of $\varepsilon_{l,\text{hBN}}$ and ω. (**b**) The relationship between the real parts of $\varepsilon_{l,\text{p-SiC}}$ and ω with $f = 0.3$. Here $l = xx, zz$ and grey shadow zones indicate the hyperbolic bands. The inset shows the real parts of $\varepsilon_{l,\text{p-SiC}}$ from 1.72 ω_0 to 1.84 ω_0.

Bulk SiC is an isotropic material, and its permittivity can be described by the Lorentz model [40]

$$\varepsilon_s(\omega) = \varepsilon_\infty \frac{\omega^2 - \omega_{\text{L}}^2 + i\gamma\omega}{\omega^2 - \omega_{\text{T}}^2 + i\gamma\omega}, \tag{3}$$

where $\varepsilon_\infty = 6.7$, $\omega_{\text{L}} = 1.827\ \omega_0$, $\omega_{\text{T}} = 1.495\ \omega_0$, and $\gamma = 0.009\ \omega_0$. The desired kind of anisotropy can be generated by the structure of a lattice of nanowires embedded in a dielectric matrix. Therefore, by embedding a lattice of air cylinders in a SiC, an artificial HM can be fabricated. By using the Maxwell-Garnett method [50], the effective permittivity of such porous SiC is described as

$$\varepsilon_{xx,\text{p-SiC}} = \varepsilon_{yy,\text{p-SiC}} = \frac{[(1+f) + (1-f)\varepsilon_s]\varepsilon_s}{(1-f) + (1+f)\varepsilon_s}, \tag{4}$$

$$\varepsilon_{zz,\text{p-SiC}} = f + (1-f)\varepsilon_s, \tag{5}$$

where the filling factor f is the area percentage occupied by air holes in the xy section of the medium. The real parts of $\varepsilon_{l,\text{p-SiC}}$ ($l = xx, zz$) as functions of ω with $f = 0.3$ are presented in Figure 2b. Two hyperbolic bands of porous SiC are obtained, as shown by the inset in Figure 2b. Artificial materials similar to such porous SiC are also named hyperbolic metamaterials that can also support HPhPs.

By utilizing the Maxwell electromagnetic stress tensor method with the properties of macroscopic field operators, the Casimir force at zero temperature is eventually expressed as [23]

$$F = -\frac{\hbar}{\pi}\text{Re}\int_0^\infty d\omega \iint \frac{d^2\mathbf{k}_\parallel}{2\pi}\sqrt{\frac{\omega^2}{c^2}-k_\parallel^2}\sum_{p=\text{TE,TM}}\frac{r_{1p}(\omega,k)r_{2p}(\omega,k)e^{2id\sqrt{\omega^2/c^2-k_\parallel^2}}}{1-r_{1p}(\omega,k)r_{2p}(\omega,k)e^{2id\sqrt{\omega^2/c^2-k_\parallel^2}}}, \tag{6}$$

where the integral is carried out over all electromagnetic modes. The wave vector component $\mathbf{k}_\parallel$ is parallel to the $x-y$ plane. The reflection coefficient from the space between two samples to the surface of top (bottom) sample for a p polarized wave is denoted by r_{1p} (r_{2p}). Since the top and bottom samples in each model are identical, r_{1p} equals r_{2p}. We shall omit the subscripts 1 and 2. All the singularities can be avoided by converting the integral of positive real ω to that of positive imaginary frequency ξ, i.e., $\omega = i\xi$, the Casimir force can be written as

$$F = \frac{\hbar}{2\pi^2}\int_0^\infty d\xi\int_0^\infty k_\parallel dk_\parallel\sqrt{\frac{\xi^2}{c^2}+k_\parallel^2}\sum_{p=\text{TE,TM}}\frac{r_p(i\xi,k)r_p(i\xi,k)e^{-2d\sqrt{\xi^2/c^2+k_\parallel^2}}}{1-r_p(i\xi,k)r_p(i\xi,k)e^{-2d\sqrt{\xi^2/c^2+k_\parallel^2}}}. \tag{7}$$

To compute the Casimir force, the reflection coefficients are obtained using the approach described in Reference [51]. Graphene is a monolayer in this study that can be considered as a conductivity current. Appendix A contains the detailed derivation of r_{TE} and r_{TM}. The reflection coefficient of the graphene-covered HM for the TE polarized wave can be written as

$$r_{\text{TE}} = \frac{k_{iz}-k_{tz}^{\text{TE}}-\sigma\omega\mu_0}{k_{iz}+k_{tz}^{\text{TE}}+\sigma\omega\mu_0}, \tag{8}$$

where $k_{iz} = \sqrt{k_0^2-k_\parallel^2}$ and $k_{tz}^{\text{TE}} = \sqrt{\varepsilon_{xx}k_0^2-k_\parallel^2}$. The wave vector in free space is $k_0 = \omega/c$. Since only ε_{xx} appears in r_{TE}, HPhPs can not be excited by TE polarized waves. The reflection coefficient for the TM polarized wave is expressed as

$$r_{\text{TM}} = \frac{\varepsilon_{xx}k_{iz}-k_{tz}^{\text{TM}}+\frac{\sigma k_{iz}k_{tz}^{\text{TM}}}{\omega\varepsilon_0}}{\varepsilon_{xx}k_{iz}+k_{tz}^{\text{TM}}+\frac{\sigma k_{iz}k_{tz}^{\text{TM}}}{\omega\varepsilon_0}}, \tag{9}$$

where $k_{tz}^{\text{TM}} = \sqrt{\varepsilon_{xx}k_0^2-k_\parallel^2\varepsilon_{xx}/\varepsilon_{zz}}$. Obviously, TM polarized waves can excite HPhPs since both ε_{xx} and ε_{zz} appear in r_{TM}. Furthermore, the reflection coefficients are affected by graphene conductivity, implying that SPs supported by graphene can couple with electromagnetic modes supported by HM, particularly HPhPs. Because the optical properties of the sample can be conveniently turned by varying E_F, the Casimir force, which is usually dependent on the surrounding environment, can be controlled by E_F.

3. Results and Discussion

3.1. Casimir Force of Graphene-Covered hBN

The relative Casimir forces between two identical graphene-covered hBN bulks as a function of separating distance d for various Fermi levels are presented in Figure 3.

The Casimir force is scaled by the well-known formula $F_0 = \hbar c\pi^2/240d^4$, which is the Casimir force per unit area between two parallel perfectly conducting plates separated by d. Because two samples have the same electric and magnetic properties, the force is obviously attractive at any distance. The relative force between two identical hBN bulks is also plotted for comparison purposes, as shown by the blue line in Figure 3. The relative force is clearly increased when hBN bulks are covered by graphene. In addition, as the Fermi level E_f increases, the relative force increases monotonically for arbitrary separating distances. The relative force is insensitive to E_f for minimal d, and curve slopes are large. However, for large d, the relative force is sensitive to E_f, but curve slopes are small. Since the Fermi level is adjustable, the Casimir force can be controlled flexibly.

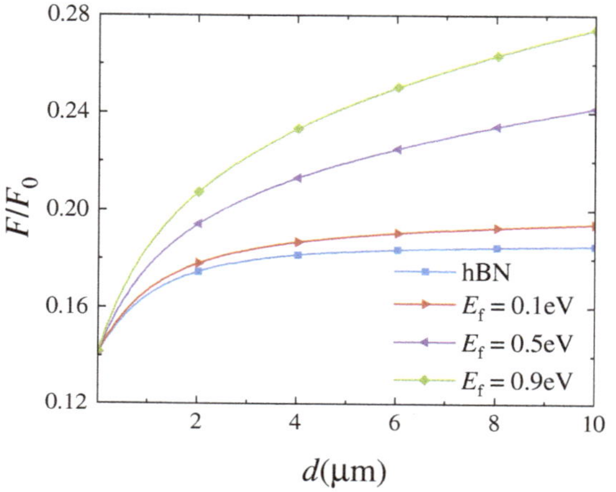

Figure 3. The relationship between the relative Casimir force and the separating distance d. The blue line represents the force between two identical hBN bulks, while the other lines represent the force between two graphene-covered hBN bulks at different Fermi levels.

The Casimir force is related to all electromagnetic modes supported by two samples. From Equation (6), we know that electromagnetic modes are represented by the reflection coefficients of samples. Thus, both real parts of r_{TM} and r_{TE} are plotted as a function of frequency ω and wave vector component $k_{||}$ in Figure 4. Comparing Re(r_{TM}) of hBN and Re(r_{TM}) of graphene-covered hBN, i.e., Figure 4a,b, reflection coefficients are enhanced clearly when $\omega < 0.1\omega_0$ for all $k_{||}$. However, it is difficult to distinguish between Re(r_{TE}) of hBN and Re(r_{TE}) of graphene-covered hBN by comparing Figure 4c,d. According to Equation (8), when hBN is covered by graphene, r_{TE} is affected by σ beyond $\varepsilon_{xx,hBN}$. By analyzing Equation (9), when the sample is graphene-covered hBN, r_{TM} is affected by $\varepsilon_{xx,hBN}$, $\varepsilon_{zz,hBN}$ and σ, whereas it only relates to the permittivity of hBN when the graphene is absent. Therefore, SPs supported by graphene are mainly coupled with HPhPs supported by hBN. As a result, the Casimir force is enhanced by covering hBN with graphene, as shown in Figure 3. Therefore, this study focuses on the coupling of SPs and HPhPs excited by a TM polarized wave.

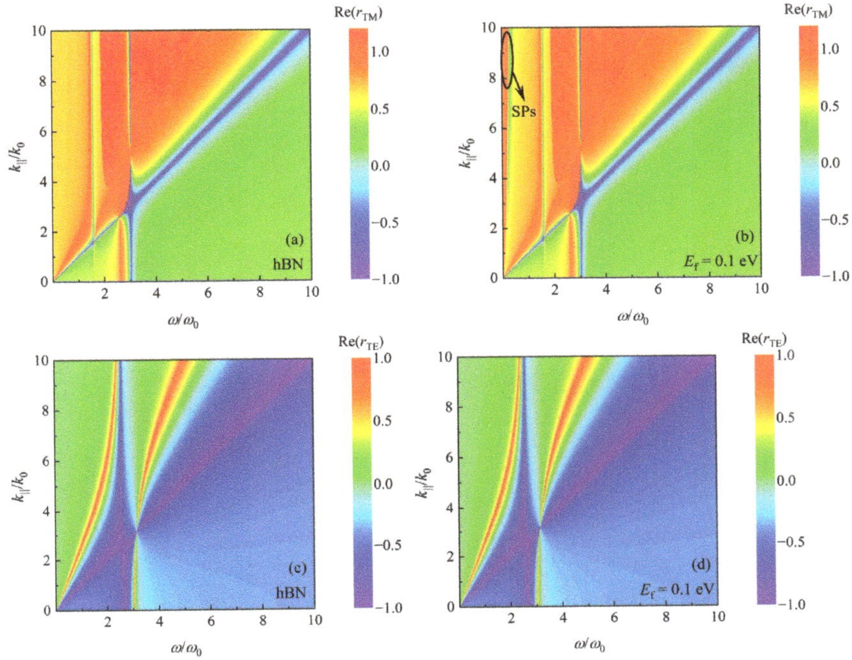

Figure 4. Re(r_{TM}) or Re(r_{TE}) as function of ω and $k_{||}$. (**a**) Re(r_{TM}) of hBN. (**b**) Re(r_{TM}) of graphene-covered hBN with Fermi level $E_f = 0.1$ eV. (**c**) Re(r_{TE}) of hBN. (**d**) Re(r_{TE}) of graphene-covered hBN with Fermi level $E_f = 0.1$ eV.

To extract the contribution of SPs and HPhPs to the enhancement of the Casimir force, the TM reflection coefficients as a function of imaginary frequency ξ and wave vector component $k_{||}$ are presented in Figure 5. When comparing Figure 5a,b, it is clear that $r_{TM}(i\xi)$ is enhanced at low frequencies when the hBN is covered by graphene, and as shown in Figure 5b–d, the enhancement area grows as the Fermi level increases. In Equation (7), the term $\exp(-2d\sqrt{\xi^2/c^2 + k_{||}^2})$ acts as a truncated function. The arc of a circle $\xi^2/c^2 + k_{||}^2 = (1/2d)^2$ for $d = 1\mu$m is plotted in Figure 5. The amplitude of the force can be represented by reflection coefficients inside the arc [23]. Clearly, the proportion of high reflection coefficients inside this arc grows as E_F increases. Correspondingly, the Casimir force is getting larger for $d = 1$ μm with increasing E_F, as shown in Figure 3. Furthermore, the radius of a circular arc is inversely proportional to the separating distance d. In Figure 5d, we also plot the curves for $d = 0.5$ μm and $d = 3$ μm. Obviously, with increasing d, i.e., decreasing radius, the proportion of high reflection coefficients is increasing. As a result, the Casimir force tends to F_0 with expanding d, as illustrated in Figure 3.

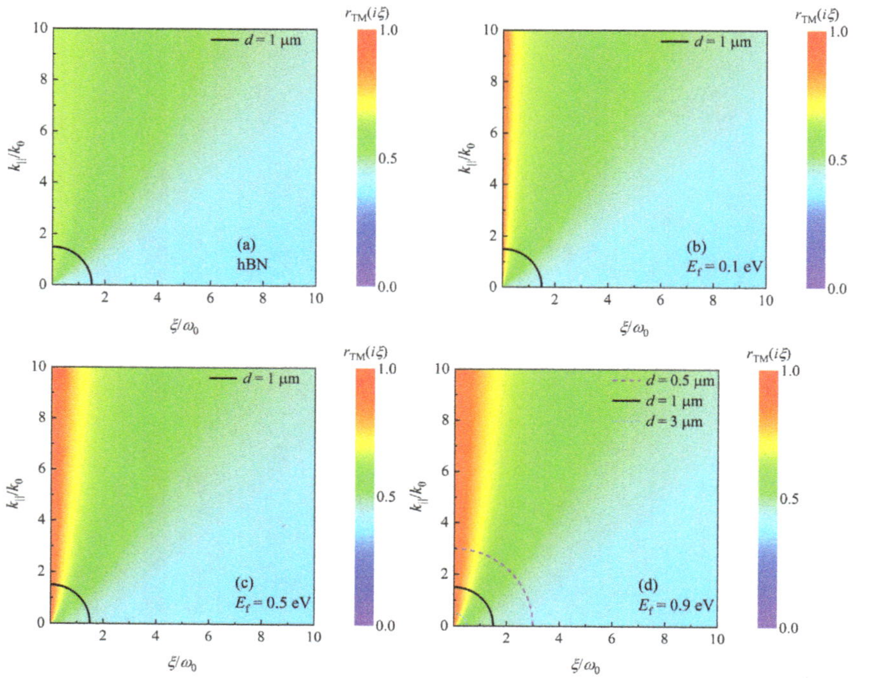

Figure 5. $r_{TM}(i\xi)$ as function of ξ and $k_{||}$. (**a**) is the case of hBN. (**b–d**) are the cases of graphene-covered hBN with Fermi levels (**b**) $E_f = 0.1$ eV, (**c**) $E_f = 0.5$ eV and (**d**) $E_f = 0.9$ eV, respectively. Solid curves in panels (**a–d**) indicate the arc of a circle $\xi^2/c^2 + k_{||}^2 = (1/2d)^2$ for $d = 1$ μm. Dashed and dotted lines in panel (**d**) indicate the cases for $d = 0.5$ μm and $d = 3$ μm, respectively.

3.2. Casimir Force of Graphene-Covered Porous SiC

The Casimir force between two identical graphene-covered porous SiC bulks is also investigated in relation to the separating distance d. The filling factor is $f = 0.1$, and the forces for varying Fermi levels are presented in Figure 6a. The force between two porous SiC bulks is plotted for comparison purposes, as shown by the blue line in Figure 6a. Similar to the case of graphene-covered hBN, the Casimir force increases monotonically with increasing Fermi levels, particularly at large separating distances. These results are expected according to the above analysis because porous SiC is also an HM. Furthermore, when the Fermi level and separating distance are fixed, a more significant force can be obtained in the graphene-covered porous SiC configuration than in the graphene-covered hBN bulks, as shown in Figure 3. As shown in Figure 2b, both $\varepsilon_{xx,\text{p-SiC}}$ and $\varepsilon_{zz,\text{p-SiC}}$ are negative in the range 1.496 ω_0 to 1.761 ω_0, indicating that porous SiC also excites electromagnetic modes other than HPhPs. These modes are all coupled and contribute to the Casimir force.

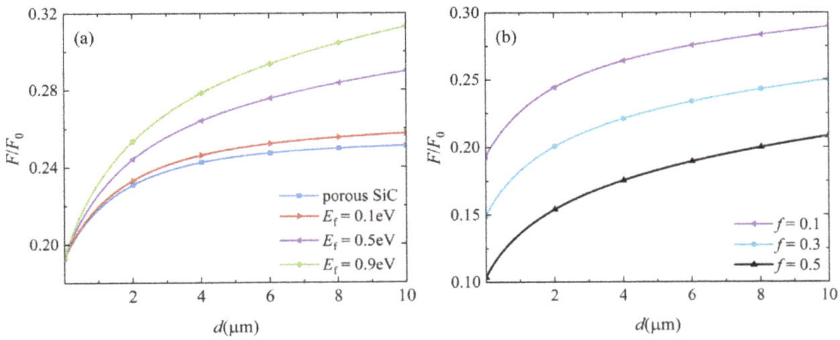

Figure 6. Dependence of the Casimir force on the separation d. (**a**) The case of identical porous SiC is indicated by the blue line, while other lines represent the case of graphene-covered porous SiC at different Fermi levels. The filling factor has been set to $f = 0.1$. (**b**) The case of graphene-covered porous SiC for different filling factors with a fixed Fermi level $E_f = 0.5$ eV.

As an artificial HM, the permittivity of porous SiC can be modulated by the filling factor f. Therefore, f influences the coupling of SPs supported by graphene and HPhPs supported by porous SiC, which can be used to control the Casimir force. Figure 6b depicts the Casimir force as a function of the separating distance d for different filling factors. When f increases, the relative force for arbitrary separations decreases dramatically. This outcome is simple to comprehend. As f increases, so does the proportion of air holes. According to Equations (4) and (5), the electric properties of porous SiC should decrease as f increases. As a result, the Casimir force decreases as the contribution of these modes supported by porous SiC decreases. Figure 7 depicts the permittivity of porous SiC as a function of ξ for different filling factors to demonstrate this explanation. As the filling factor f increases, for arbitrary ξ, both $\varepsilon_{xx,\text{p-SiC}}$ and $\varepsilon_{zz,\text{p-SiC}}$ decrease, confirming our prediction. Furthermore, when $\xi < \omega_0$, all curves in Figure 7 are almost flat, but sloping when $\omega_0 < \xi < 10\omega_0$. As previously stated, the force at a small separation distance is primarily derived from electromagnetic modes in the high-frequency region, whereas modes in the low-frequency region mainly contribute to the force at a large separation distance [22]. As shown in Figure 6b, slopes of relative forces at a small separation distance are large, while slopes at a large separation distance are small.

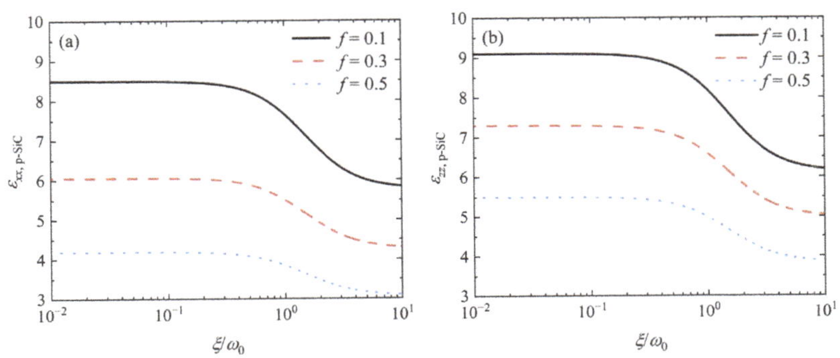

Figure 7. (**a**) $\varepsilon_{xx,\text{p-SiC}}$ and (**b**) $\varepsilon_{zz,\text{p-SiC}}$ as a function of ξ for different filling factors f.

3.3. Discussion

The Casimir force per unit area between two parallel graphene sheets is inversely proportional to d^4 at zero temperature, but with a substantially smaller coefficient when compared with that of two perfectly conducting plates [38]. According to the data in Reference [38], the Casimir force per 1cm^2 is around 0.006 $F_0 = 7.8 \times 10^{-6}$ N when

$d = 0.1$ μm. For the same area and d, when $E_f = 0.1$ eV, F is around $0.145\ F_0 = 1.9 \times 10^{-4}$ N in Figure 3, and $0.197\ F_0 = 2.6 \times 10^{-4}$ N in Figure 6a. Therefore, mounting graphene on an HM substrate increases the Casimir force by more than one order of magnitude. Furthermore, the Casimir force per 1 cm^2 between two identical artificial HMs is around $0.1\ F_0 = 8.1 \times 10^{-10}$ N at $d \approx 2$ μm [23]. To compare, when $E_f = 0.1$ eV, F is about $0.178\ F_0 = 1.5 \times 10^{-9}$ N in Figure 3, and $0.233\ F_0 = 1.9 \times 10^{-9}$ N in Figure 6a. Covering an HM with graphene monolayer therefore enhances the Casimir force.

Since the samples in this study are in free space, the detection system should be set in a high-vacuum chamber to measure the Casimir force between graphene-covered HMs. Furthermore, the quality of the graphene layer should be checked to guarantee that it is a monolayer. The gradient of the Casimir force may be measured using the experimental setup described in Reference [7]. However, there are some problems. One difficulty is that maintaining two flat samples parallel over small distances is quite challenging. Most experiments chose one of the samples to be spherical with a large radius. As a result, the geometry will influence the magnitude of the Casimir force. Another issue is the effect of finite temperature since the temperature considered here is zero. The Casimir force at finite temperature can be obtained by substituting the integration along with the imaginary frequency ξ axis in Equation (7) with the summation over the Matsubara frequencies [24]. The influences of geometry and finite temperature on the Casimir force of our proposal are yet to be investigated further.

4. Conclusions

In conclusion, we investigate the Casimir force between two identical graphene-covered HMs. The first model's samples are graphene-covered hBN bulks. When graphene covers hBN, the Casimir force increases for arbitrary separation distances, and as the Fermi level E_f increases, the force increases monotonically. The reflection coefficients of samples are thoroughly examined. SPs supported by graphene and HPhPs supported by hBN are coupled, and these electromagnetic modes are related to the enhancement of the Casimir force. Furthermore, the hBN is replaced by porous SiC, which is treated as an artificial HM following the effective medium theory. When E_f increases, the Casimir force is still increased monotonically. However, as the filling factor f increases, the force decreases for arbitrary separation distances. The electromagnetic responses of porous SiC are used to understand this phenomenon. As a result, by designing suitable artificial HMs and tuning the Fermi level E_f, the desired Casimir force between graphene-covered HMs can be controlled. By combining graphene and HM, this work provides a flexible way to modulate the Casimir effect, which can use for the detection of Casimir force.

Author Contributions: Conceptualization, G.S.; methodology, G.S.; validation, G.S., Z.L. and L.J.; formal analysis, G.S., C.L. and Y.C.; investigation, G.S., C.L. and Y.C.; writing—original draft preparation, G.S.; writing—review and editing, G.S., Z.L. and L.J.; visualization, G.S.; funding acquisition, G.S. and Z.L. All authors have read and agreed to the published version of the manuscript.

Funding: This research was funded by the Fishery Engineering and Equipment Innovation Team of Shanghai High-level Local University.

Institutional Review Board Statement: Not applicable.

Informed Consent Statement: Not applicable.

Data Availability Statement: The data presented in this study are available on request from the corresponding author.

Conflicts of Interest: The authors declare no conflict of interest.

Appendix A

Figure A1 depicts a schematic of electromagnetic wave scattering at the interface of vacuum and a graphene-covered HM. Assuming graphene is located in $z = 0$. The medium in $z < 0$ is vacuum, whereas the HM occupies $z > 0$ space. An electromagnetic wave incident from the vacuum into the graphene-covered HM produces a reflected field in the vacuum and a transmitted field in the HM. Because graphene is a monolayer, it can be treated as a conducting sheet overlaying on an HM with infinite thickness in this work. We consider the TM polarized wave, and the magnetic fields of the incident, reflected, and transmitted waves can be written as

$$\vec{H}_{\rm i} = Ae^{i[(k_{||}x+k_{iz}z)-\omega t]}\vec{j} \tag{A1}$$

$$\vec{H}_{\rm r} = Re^{i[(k_{||}x-k_{iz}z)-\omega t]}\vec{j} \tag{A2}$$

$$\vec{H}_{\rm t} = Te^{i[(k_{||}x+k_{tz}^{\rm TM}z)-\omega t]}\vec{j} \tag{A3}$$

where A, R, and T denote the magnetic field magnitudes of the incident, reflected, and transmitted fields, respectively. The electric fields of the incident, reflected, and transmitted waves can be expressed using Maxwell's equations as

$$\vec{E}_{\rm i} = \frac{A}{\omega\varepsilon_0}e^{i[(k_{||}x+k_{iz}z)-\omega t]}(k_{iz}\vec{i} - k_{||}\vec{k}) \tag{A4}$$

$$\vec{E}_{\rm r} = \frac{R}{\omega\varepsilon_0}e^{i[(k_{||}x-k_{iz}z)-\omega t]}(-k_{iz}\vec{i} - k_{||}\vec{k}) \tag{A5}$$

$$\vec{E}_{\rm t} = \frac{T}{\omega\varepsilon_0}e^{i[(k_{||}x+k_{tz}^{\rm TM}z)-\omega t]}(\frac{k_{tz}^{\rm TM}}{\varepsilon_{xx}}\vec{i} - \frac{k_{||}}{\varepsilon_{zz}}\vec{k}) \tag{A6}$$

where ε_{xx} and ε_{zz} are diagonal elements of the permittivity tensor of a uniaxial HM. $k_{||}^2 + k_{iz}^2 = k_0^2$ and $k_{||}^2/\varepsilon_{zz} + (k_{tz}^{\rm TM})^2/\varepsilon_{xx} = k_0^2$ are satisfied. Since the graphene layer is treated as a conducting sheet, the boundary conditions are expressed as

$$\vec{k} \times [\vec{E}_{\rm t} - (\vec{E}_{\rm i} + \vec{E}_{\rm r})] = 0 \tag{A7}$$

$$\vec{k} \times [\vec{H}_t - (\vec{H}_{\rm i} + \vec{H}_{\rm r})] = \vec{J} \tag{A8}$$

where $\vec{J}$ is the conductivity current. The current $\vec{J}$ contains just the x component since the electric field $\vec{E}$ is in the xz plane. By inserting Equations (A1)–(A6) into Equations (A7) and (A8), we get

$$\frac{T}{\omega\varepsilon_0}\frac{k_{tz}^{\rm TM}}{\varepsilon_{xx}} = \frac{A}{\omega\varepsilon_0}\frac{k_{iz}}{\varepsilon_1} - \frac{R}{\omega\varepsilon_0}\frac{k_{iz}}{\varepsilon_1} \tag{A9}$$

$$-[T-(A+R)] = \sigma\frac{T}{\omega\varepsilon_0}\frac{k_{tz}^{\rm TM}}{\varepsilon_{xx}} \tag{A10}$$

The reflection coefficient for the TM polarized wave can be derived and expressed as

$$r_{\rm TM} = \frac{R}{A} = \frac{\varepsilon_{xx}k_{iz} - k_{tz}^{\rm TM} + \frac{\sigma k_{iz}k_{tz}^{\rm TM}}{\omega\varepsilon_0}}{\varepsilon_{xx}k_{iz} + k_{tz}^{\rm TM} + \frac{\sigma k_{iz}k_{tz}^{\rm TM}}{\omega\varepsilon_0}} \tag{A11}$$

If the incident wave is TE polarized, the incident, reflected, and transmitted electric fields can be represented as

$$\vec{E}_{\rm i} = Ae^{i[(k_{||}x+k_{iz}z)-\omega t]}\vec{j} \tag{A12}$$

$$\vec{E}_{\rm r} = Re^{i[(k_{||}x-k_{iz}z)-\omega t]}\vec{j} \tag{A13}$$

$$\vec{E}_{\rm t} = Te^{i[(k_{||}x+k_{tz}^{\rm TE}z)-\omega t]}\vec{j} \tag{A14}$$

where A, R, and T represent the electric field magnitudes of the incident, reflected, and transmitted waves, respectively. $k_{||}^2 + (k_{tz}^{\mathrm{TE}})^2 = \varepsilon_{xx} k_0^2$ is satisfied. The magnetic fields can then be expressed as

$$\vec{H}_{\mathrm{i}} = \frac{A}{\omega\mu_0} e^{i[(k_{||}x + k_{iz}z) - \omega t]} (-k_{iz}\vec{i} + k_{||}\vec{k}) \tag{A15}$$

$$\vec{H}_{\mathrm{r}} = \frac{R}{\omega\mu_0} e^{i[(k_{||}x - k_{iz}z) - \omega t]} (k_{iz}\vec{i} + k_{||}\vec{k}) \tag{A16}$$

$$\vec{H}_{\mathrm{t}} = \frac{T}{\omega\mu_0} e^{i[(k_{||}x + k_{tz}^{\mathrm{TE}}z) - \omega t]} (-k_{tz}^{\mathrm{TE}}\vec{i} + k_{||}\vec{k}) \tag{A17}$$

Since we are considering a TE polarized wave, the current $\vec{J}$ has just the y component. Substituting Equations (A12)–(A17) into Equations (A7) and (A8), we get

$$T - (A + R) = 0 \tag{A18}$$

$$-\frac{T}{\omega\mu_0} k_{tz}^{\mathrm{TE}} - (-\frac{A}{\omega\mu_0} k_{iz} + \frac{R}{\omega\mu_0} k_{iz}) = \sigma T \tag{A19}$$

The reflection coefficient for the TE polarized wave can be derived and expressed as

$$r_{\mathrm{TE}} = \frac{R}{A} = \frac{k_{iz} - k_{tz}^{\mathrm{TE}} - \sigma\omega\mu_0}{k_{iz} + k_{tz}^{\mathrm{TE}} + \sigma\omega\mu_0} \tag{A20}$$

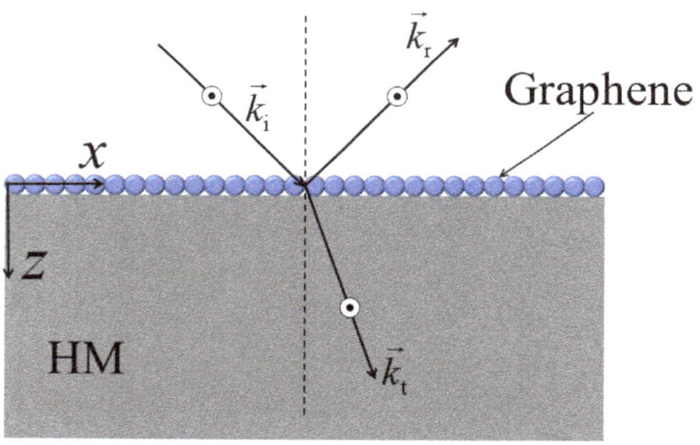

Figure A1. Schematic of electromagnetic wave scattering at the interface of vacuum and a graphene-covered HM.

References

1. Casimir, H.B. On the attraction between two perfectly conducting plates. *Proc. K. Ned. Akad. Wet.* **1948**, *51*, 793–795.
2. Lifshitz, E.M. The theory of molecular attractive forces between solids. *Soviet Phys. JETP* **1956**, *2*, 73–83.
3. Klimchitskaya, G.L.; Mostepanenko, V.M. Experiment and theory in the Casimir effect. *Contemp. Phys.* **2006**, *47*, 131–144. [CrossRef]
4. Chen, F.; Klimchitskaya, G.L.; Mostepanenko, V.M.; Mohideen, U. Control of the Casimir force by the modification of dielectric properties with light. *Phys. Rev. B* **2007**, *76*, 035338. [CrossRef]
5. Klimchitskaya, G.L.; Mohideen, U.; Mostepanenko, V.M. The Casimir force between real materials: Experiment and theory. *Rev. Mod. Phys.* **2009**, *81*, 1827–1885. [CrossRef]
6. Woods, L.M.; Dalvit, D.A.R.; Tkatchenko, A.; Rodriguez-Lopez, P.; Rodriguez, A.W.; Podgornik, R. Materials perspective on Casimir and van der Waals interactions. *Rev. Mod. Phys.* **2016**, *88*, 045003. [CrossRef]
7. Chang, C.C.; Banishev, A.A.; Castillo-Garza, R.; Klimchitskaya, G.L.; Mostepanenko, V.M.; Mohideen, U. Gradient of the Casimir force between Au surfaces of a sphere and a plate measured using an atomic force microscope in a frequency-shift technique. *Phys. Rev. B* **2012**, *85*, 165443. [CrossRef]

8. Banishev, A.A.; Chang, C.C.; Klimchitskaya, G.L.; Mostepanenko, V.M.; Mohideen, U. Measurement of the gradient of the Casimir force between a nonmagnetic gold sphere and a magnetic nickel plate. *Phys. Rev. B* **2012**, *85*, 195422. [CrossRef]
9. Banishev, A.A.; Klimchitskaya, G.L.; Mostepanenko, V.M.; Mohideen, U. Demonstration of the Casimir Force between Ferromagnetic Surfaces of a Ni-Coated Sphere and a Ni-Coated Plate. *Phys. Rev. Lett.* **2013**, *110*, 137401. [CrossRef]
10. Bimonte, G.; López, D.; Decca, R.S. Isoelectronic determination of the thermal Casimir force. *Phys. Rev. B* **2016**, *93*, 184434. [CrossRef]
11. Xu, J.; Klimchitskaya, G.L.; Mostepanenko, V.M.; Mohideen, U. Reducing detrimental electrostatic effects in Casimir-force measurements and Casimir-force-based microdevices. *Phys. Rev. A* **2018**, *97*, 032501. [CrossRef]
12. Liu, M.; Xu, J.; Klimchitskaya, G.L.; Mostepanenko, V.M.; Mohideen, U. Precision measurements of the gradient of the Casimir force between ultraclean metallic surfaces at larger separations. *Phys. Rev. A* **2019**, *100*, 052511. [CrossRef]
13. Liu, M.; Xu, J.; Klimchitskaya, G.L.; Mostepanenko, V.M.; Mohideen, U. Examining the Casimir puzzle with an upgraded AFM-based technique and advanced surface cleaning. *Phys. Rev. B* **2019**, *100*, 081406. [CrossRef]
14. Serry, F.M.; Walliser, D.; Maclay, G.J. The role of the casimir effect in the static deflection and stiction of membrane strips in microelectromechanical systems (MEMS). *J. Appl. Phys.* **1998**, *84*, 2501–2506. [CrossRef]
15. Buks, E.; Roukes, M.L. Stiction, adhesion energy, and the Casimir effect in micromechanical systems. *Phys. Rev. B* **2001**, *63*, 033402. [CrossRef]
16. Rodriguez, A.W.; Woolf, D.; Hui, P.C.; Iwase, E.; McCauley, A.P.; Capasso, F.; Loncar, M.; Johnson, S.G. Designing evanescent optical interactions to control the expression of Casimir forces in optomechanical structures. *Appl. Phys. Lett.* **2011**, *98*, 194105. [CrossRef]
17. Kenneth, O.; Klich, I. Opposites Attract: A Theorem about the Casimir Force. *Phys. Rev. Lett.* **2006**, *97*, 160401. [CrossRef]
18. Buks, E.; Roukes, M.L. Casimir force changes sign. *Nature* **2002**, *419*, 119–120. [CrossRef]
19. Kenneth, O.; Klich, I.; Mann, A.; Revzen, M. Repulsive Casimir Forces. *Phys. Rev. Lett.* **2002**, *89*, 033001. [CrossRef]
20. Schurig, D.; Mock, J.J.; Justice, B.J.; Cummer, S.A.; Pendry, J.B.; Starr, A.F.; Smith, D.R. Metamaterial Electromagnetic Cloak at Microwave Frequencies. *Science* **2006**, *314*, 977–980. [CrossRef]
21. Fan, J.W.; Xu, J.; Cheng, M.T.; Yang, Y. Vacuum induced transparency in metamaterials. *Opt. Express* **2018**, *26*, 19498–19512. [CrossRef]
22. Yang, Y.; Zeng, R.; Chen, H.; Zhu, S.; Zubairy, M.S. Controlling the Casimir force via the electromagnetic properties of materials. *Phys. Rev. A* **2010**, *81*, 022114. [CrossRef]
23. Song, G.; Xu, J.; Zhu, C.; He, P.; Yang, Y.; Zhu, S.Y. Casimir force between hyperbolic metamaterials. *Phys. Rev. A* **2017**, *95*, 023814. [CrossRef]
24. Zeng, R.; Yang, Y.; Zhu, S. Casimir force between anisotropic single-negative metamaterials. *Phys. Rev. A* **2013**, *87*, 063823. [CrossRef]
25. Zeng, R.; Yang, Y. Tunable polarity of the Casimir force based on saturated ferrites. *Phys. Rev. A* **2011**, *83*, 012517. [CrossRef]
26. Phan, A.D.; Viet, N.A.; Poklonski, N.A.; Woods, L.M.; Le, C.H. Interaction of a graphene sheet with a ferromagnetic metal plate. *Phys. Rev. B* **2012**, *86*, 155419. [CrossRef]
27. Grushin, A.G.; Cortijo, A. Tunable Casimir Repulsion with Three-Dimensional Topological Insulators. *Phys. Rev. Lett.* **2011**, *106*, 020403. [CrossRef]
28. Nie, W.; Zeng, R.; Lan, Y.; Zhu, S. Casimir force between topological insulator slabs. *Phys. Rev. B* **2013**, *88*, 085421. [CrossRef]
29. Zeng, R.; Chen, L.; Nie, W.; Bi, M.; Yang, Y.; Zhu, S. Enhancing Casimir repulsion via topological insulator multilayers. *Phys. Lett. A* **2016**, *380*, 2861–2869. [CrossRef]
30. Geim, A.K.; Novoselov, K.S. The rise of graphene. *Nat. Mater.* **2007**, *6*, 183–191. [CrossRef]
31. Castro Neto, A.H.; Guinea, F.; Peres, N.M.R.; Novoselov, K.S.; Geim, A.K. The electronic properties of graphene. *Rev. Mod. Phys.* **2009**, *81*, 109–162. [CrossRef]
32. Yu, T.; Luo, R.; Wang, T.; Zhang, D.; Liu, W.; Yu, T.; Liao, Q. Enhancement of Casimir Friction between Graphene-Covered Topological Insulator. *Nanomaterials* **2022**, *12*, 1148. [CrossRef]
33. Luo, R.; Yang, J.R.; Wang, T.B.; Zhang, D.J.; Liu, W.X.; Yu, T.B.; Liao, Q.H. Enhancement of Casimir friction between graphene-covered hyperbolic materials. *Phys. Lett. A* **2021**, *387*, 127006. [CrossRef]
34. Fialkovsky, I.V.; Marachevsky, V.N.; Vassilevich, D.V. Finite-temperature Casimir effect for graphene. *Phys. Rev. B* **2011**, *84*, 035446. [CrossRef]
35. Klimchitskaya, G.L.; Mostepanenko, V.M. van der Waals and Casimir interactions between two graphene sheets. *Phys. Rev. B* **2013**, *87*, 075439. [CrossRef]
36. Klimchitskaya, G.L.; Mostepanenko, V.M.; Sernelius, B.E. Two approaches for describing the Casimir interaction in graphene: Density-density correlation function versus polarization tensor. *Phys. Rev. B* **2014**, *89*, 125407. [CrossRef]
37. Klimchitskaya, G.L.; Mostepanenko, V.M. Classical limit of the Casimir interaction for thin films with applications to graphene. *Phys. Rev. B* **2014**, *89*, 035407. [CrossRef]
38. Drosdoff, D.; Woods, L.M. Casimir interactions between graphene sheets and metamaterials. *Phys. Rev. A* **2011**, *84*, 062501. [CrossRef]
39. Klimchitskaya, G.L.; Mostepanenko, V.M. Casimir and Casimir-Polder Forces in Graphene Systems: Quantum Field Theoretical Description and Thermodynamics. *Universe* **2020**, *6*, 150. [CrossRef]

40. Wang, T.B.; Zhou, Y.; Mu, H.Q.; Shehzad, K.; Zhang, D.J.; Liu, W.X.; Yu, T.B.; Liao, Q.H. Enhancement of lateral Casimir force on a rotating particle near hyperbolic metamaterial. *Nanotechnology* **2022**, *33*, 245001. [CrossRef]
41. Novoselov, K.S.; Fal'ko, V.I.; Colombo, L.; Gellert, P.R.; Schwab, M.G.; Kim, K. A roadmap for graphene. *Nature* **2012**, *490*, 192–200. [CrossRef]
42. Zeng, X.; Al-Amri, M.; Zubairy, M.S. Tunable Goos-Hänchen shift from graphene ribbon array. *Opt. Express* **2017**, *25*, 23579–23588. [CrossRef]
43. Din, R.U.; Zeng, X.; Ahmad, I.; Ge, G.Q. Enhanced and highly tunable Goos-Hänchan shifts at a nanocomposite-graphene interface. *Appl. Phys. Lett.* **2019**, *114*, 161902. [CrossRef]
44. Zeng, X.; Li, Z.; Ge, G.Q.; Zubairy, M.S. Quantum interference near graphene layers: Observing the surface plasmons with transverse electric polarization. *Phys. Rev. A* **2019**, *99*, 043811. [CrossRef]
45. Poddubny, A.; Iorsh, I.; Belov, P.; Kivshar, Y. Hyperbolic metamaterials. *Nat. Photonics* **2013**, *7*, 948–957. [CrossRef]
46. Shekhar, P.; Atkinson, J.; Jacob, Z. Hyperbolic metamaterials: fundamentals and applications. *Nano Converg.* **2014**, *1*, 1–17. [CrossRef]
47. Lin, Y.; Connell, J.W. Advances in 2D boron nitride nanostructures: nanosheets, nanoribbons, nanomeshes, and hybrids with graphene. *Nanoscale* **2012**, *4*, 6908–6939. [CrossRef]
48. Su, C.; Fu, C. Surface and volume phonon polaritons in a uniaxial hyperbolic material: optic axis parallel versus perpendicular to the surface. *Opt. Express* **2021**, *29*, 39824–39837. [CrossRef]
49. Jacob, Z.; Smolyaninov, I.I.; Narimanov, E.E. Broadband Purcell effect: Radiative decay engineering with metamaterials. *Appl. Phys. Lett.* **2012**, *100*, 181105. [CrossRef]
50. Markel, V.A. Introduction to the Maxwell Garnett approximation: tutorial. *J. Opt. Soc. Am. A* **2016**, *33*, 1244–1256. [CrossRef]
51. Song, J.; Cheng, Q. Near-field radiative heat transfer between graphene and anisotropic magneto-dielectric hyperbolic metamaterials. *Phys. Rev. B* **2016**, *94*, 125419. [CrossRef]

nanomaterials

MDPI

Article

Enhancement of Casimir Friction between Graphene-Covered Topological Insulator

Ting Yu, Rong Luo, Tongbiao Wang *, Dejian Zhang, Wenxing Liu, Tianbao Yu and Qinghua Liao

Department of Physics, Nanchang University, Nanchang 330031, China; Yutingncu157@163.com (T.Y.); luorong_jjvc@163.com (R.L.); zhangdejian2015@163.com (D.Z.); liuwenxing@ncu.edu.cn (W.L.); yutianbao@ncu.edu.cn (T.Y.); lqhua@ncu.edu.cn (Q.L.)
* Correspondence: tbwang@ncu.edu.cn

Abstract: Casimir friction is theoretically studied between graphene-covered undoped bismuth selenide (Bi_2Se_3) in detail. In the graphene/Bi_2Se_3 composite structure, the coupling of the hyperbolic phonon polaritons supported by Bi_2Se_3 with the surface plasmons supported by graphene can lead to the hybrid surface plasmon–phonon polaritons (SPPPs). Compared with that between undoped Bi_2Se_3, Casimir friction can be enhanced by more than one order of magnitude due to the contribution of SPPPs. It is found that the chemical potential that can be used to modulate the optical characteristic of SPPPs plays an important role in Casimir friction. In addition, the Casimir friction between doped Bi_2Se_3 is also studied. The friction coefficient between doped Bi_2Se_3 can even be larger than that between graphene-covered undoped Bi_2Se_3 for suitable chemical potential due to the contribution of unusual electron surface states. The results obtained in this work are not only beneficial to the study of Casimir frictions but also extend the research ranges of topological insulators.

Keywords: casimir friction; graphene; topological insulator

1. Introduction

In nature, friction is a very common phenomenon, which extensively exists in the macroscopic to the microscopic world. However, the physical mechanism of friction in the microscopic world is different from that of the friction observed in our daily lives. In recent years, people have paid much attention to nanotribology with the rapid development of nanofabrication technology. Interestingly, there is one type of friction that appears between bodies in relative motion even without direct contact at the nanoscale, which is usually called noncontact friction [1–7]. Noncontact friction has been observed experimentally by atomic force microscopy [1,5,6] and has important practical significance in ultra-sensitive force detection. In the past decade, another noncontact friction originating from the momentum exchange of Doppler-shifted photons has attracted much attention [8]. The essence of this friction is closely related to the Casimir effect, so it is called Casimir friction [8,9]. Even at zero temperature, Casimir friction can still exist because of the vacuum fluctuation, so it will be denoted as quantum friction. Casimir frictions have been extensively studied in several configurations that are in relative motion, such as atom and atom [10], atom and plate [11–13], and plate and plate [14].

The theoretical method of Casimir friction between two objects separated by a small gap was firstly derived by Pendry with the help of classical electromagnetic theory [8]. Then, Volokitin and Persson also made a series of works on Casimir friction [15–20]. Because the Casimir friction is extremely small, it makes a great challenge to detect in an experiment. Enhancing the Casimir friction between bodies in relative motion becomes an important topic. An enhancement mechanism of Casimir friction associated with resonant photon tunneling on different surfaces was proposed [16]. It has been found that Casimir friction will be significantly enhanced if the materials can support low-frequency surface plasmons (SPs) or other surface polaritons [16]. Therefore, plenty of materials that can support surface polaritons are employed to investigate the Casimir friction [21,22]. Among them, graphene is a two-dimensional material composed of carbon atoms [23–25], which can support SPs from terahertz (THz) to infrared frequency ranges [26–35]. In particular, the optical conductivity of graphene depends on its chemical potential, which is controlled by an external field or gate voltage. It is shown that SPs supported by graphene can play an important role in enhancing and actively modulating Casimir friction [36–39]. In addition, the Casimir friction between graphene-covered hyperbolic materials (HMs) has also been studied [40]. The coupling of graphene plasmon with hyperbolic phonon polaritons (HPPs) supported by HMs can enhance the Casimir friction remarkably [40].

In the past decade, topological insulators have received extensive attention because of their exotic characteristic and important applications in the fabrications of new electronic devices [41–45]. They are characterized by a full insulating gap in the bulk and gapless surface states protected by time-reversal symmetry [44]. The three-dimensional (3D) topological insulators have been predicted theoretically in the $Bi_{1-x}Sb_x$ alloy [46] and observed experimentally by angle-resolved photoemission spectroscopy [47]. Then, several simple 3D topological insulators were predicted theoretically in Bi_2Te_3, Sb_2Te_3 [48], and Bi_2Se_3 [48,49] compounds with a large gap in bulk and gapless surface state. Bismuth-based topological insulators have attracted great interest due to their unusual electronic surface states, which are manifested as massless Dirac fermions [43,44,48–51]. As one of the representative topological insulators, Bi_2Se_3 is also well-known for its bulk optical response [52–55] besides its novel surface state. It exhibits hyperbolic dispersion relation and can support highly oriented collective HPPs in the THz range [56]. For Bi_2Se_3 with finite doping, the coupling of HPPs with Dirac plasmon confined at the surface can result in the appearance of hybrid modes [56]. However, the SPs are absent for undoped Bi_2Se_3. As has been mentioned above, graphene plasmon can be coupled with different surface polaritons, so it is interesting to investigate the coupling between graphene plasmon and HPPs supported by undoped Bi_2Se_3.

In this paper, we study the Casimir friction between graphene-covered undoped Bi_2Se_3. The low-frequency HPPs supported by Bi_2Se_3 can be coupled with SPs supported by graphene, and then resulting in the appearance of hybrid surface plasmon–phonon polaritons (SPPPs). We demonstrate that the SPPPs play an important role in the Casimir friction between graphene-covered Bi_2Se_3. Compared with that between undoped Bi_2Se_3, the Casimir friction between graphene-covered Bi_2Se_3 enhances about one order of magnitude in wide separation gaps. Furthermore, the hybrid SPPPs can be flexibly controlled by tuning the chemical potential of graphene that depends on the gate voltage. Therefore, it provides an opportunity to actively modulate the Casimir friction between topological insulators. Casimir friction between doped Bi_2Se_3 is also studied. Depending on the chemical potential, Casimir friction between doped Bi_2Se_3 can be larger or smaller than that between graphene-covered Bi_2Se_3 under the same separation distance. This study not only extends the research ranges of topological insulators but also provides an efficient method to control the Casimir friction.

2. Theoretical Model

The schematic of the Casimir force between graphene-covered Bi_2Se_3 is shown in Figure 1. Two graphene-covered Bi_2Se_3 bulks are separated by a distance d. We assume the top graphene-covered Bi_2Se_3 moves relatively to the bottom one along the x-axis with a velocity v in the laboratory frame. The temperature of the surrounding environment is denoted by T, which is set to be $T = 300$ K in all the calculations. If the moving velocity of the top graphene-covered Bi_2Se_3 satisfies the conditions $v < dk_BT/\hbar$, Casimir friction is proportional to the sliding velocity, which can be determined as $f = \gamma v$ [18]. k_B and $\hbar$ are the Boltzmann and reduced Planck constants, respectively. $\gamma = \gamma^{\text{rad}} + \gamma^{\text{evan}}$ is the Casimir friction coefficient, which comes from the contributions of propagation and evanescent electromagnetic waves from the graphene-covered Bi_2Se_3. However, the contribution of the propagation wave can be ignored in the near-field region [2,18]. Therefore, we only consider the Casimir friction coefficient from the evanescent waves, which can be expressed as [18]

$$\gamma \approx \frac{\hbar}{2\pi^2} \int_0^\infty d\omega \left(-\frac{\partial n}{\partial \omega}\right) \int_{\omega/c}^\infty dq q^3 e^{-2|k_z|d} \frac{\mathrm{Im}(R_{1p})\mathrm{Im}(R_{2p})}{\left|1 - e^{-2|k_z|d} R_{1p} R_{2p}\right|^2} + [p \to s] \tag{1}$$

where $n(\omega) = 1/\left(e^{\hbar\omega/k_B T} - 1\right)$ is the Bose–Einstein factor. q and $k_z = \sqrt{(\omega/c)^2 - q^2}$ are the components of the wave vector components parallel and perpendicular to the xy plane, respectively. c is the speed of light in the vacuum and R_{ip} (i = 1,2) is the reflection amplitude from the top or bottom surface of the p-polarized electromagnetic waves. The symbol $[p \to s]$ represents the p-polarized reflection amplitude R_p replaced with s-polarized reflection amplitude R_s. As described in Ref. [57], p-polarized wave often plays a dominant role in the near-field region, so the contribution from s-polarized waves can be excluded. The integral of q in Equation (1) is defined as force spectral density (FSD), which can be expressed by

$$f_p = -\frac{\partial n}{\partial \omega} \int_{\omega/c}^\infty dq q^3 \frac{\mathrm{Im}(R_{1p})\mathrm{Im}(R_{2p})}{\left|1 - e^{-2|k_z|d} R_{1p} R_{2p}\right|^2} e^{-2|k_z|d} \tag{2}$$

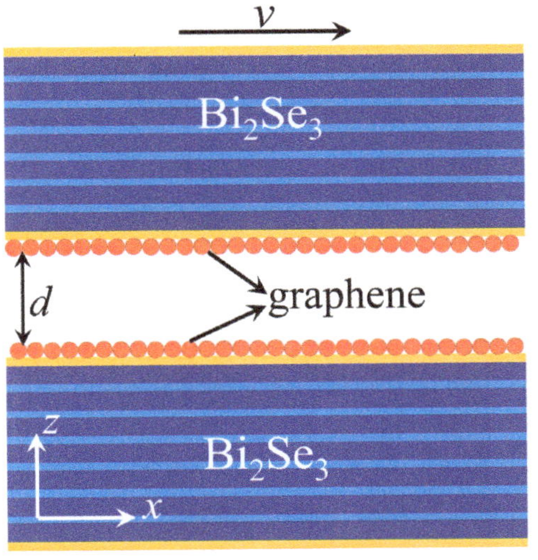

Figure 1. Schematic of Casimir friction between graphene-covered undoped Bi_2Se_3 separated by a distance d. The layered structure denotes Bi_2Se_3 and the two thin orange layers at the top and bottom of Bi_2Se_3 represent the surface states.

For convenient description, we define the photon exchange function as

$$\xi_p = \frac{\mathrm{Im}(R_{1p})\mathrm{Im}(R_{2p})}{\left|1 - e^{-2|k_z|d} R_{1p} R_{2p}\right|^2} e^{-2|k_z|d} \tag{3}$$

which is used to describe the exchange ability of photons between the bodies in relative motion. The reflection amplitude of graphene-covered anisotropic material can be written as [58]

$$R_p = \frac{p\varepsilon^{\perp} - p_p + \mu_0 c\sigma p p_p}{p\varepsilon^{\perp} + p_p + \mu_0 c\sigma p p_p} \tag{4}$$

where $p = \sqrt{1-\kappa^2}$, $p_p = \sqrt{\varepsilon^{\perp} - \varepsilon^{\perp}\kappa^2/\varepsilon^z}$, and $\kappa = cq/\omega$. ε^z and $\varepsilon^{\perp}$ are the permittivities of Bi_2Se_3 parallel and perpendicular to the optical axis (z-axis), respectively. μ_0 is the permeability in the vacuum. $\sigma(\omega)$ is the optical conductivity of graphene, which can be simplified to $\sigma = \frac{ie^2\mu}{\pi\hbar^2(\omega+i\tau^{-1})}$ in the low-frequency range by ignoring the contribution of interband transitions [32]. e is the electron charge, μ is the chemical potential of graphene, and $\tau = 10^{-13}$ s is the relaxation time. As a naturally anisotropic material, Bi_2Se_3 exhibits hyperbolic optical properties. The dielectric properties perpendicular ($\varepsilon^{\perp}$) and parallel (ε^z) to the optical axis have different values which are given by [59,60]

$$\varepsilon^{\alpha}(\omega) = \varepsilon^{\alpha}_{\infty} + \sum_{j=1,2} \frac{\omega^{\alpha\,2}_{p,j}}{\omega^{\alpha\,2}_{to,j} - \omega^2 - i\gamma^{\alpha}_{j}\omega}, (\alpha = \perp, z) \quad (5)$$

The parameters used in Equation (5) are $\varepsilon^{\perp}_{\infty} = 29$, $\varepsilon^{z}_{\infty} = 17.4$, $\omega^{\perp}_{to,1} = 64\ \text{cm}^{-1}$, $\omega^{\perp}_{p,1} = 704\ \text{cm}^{-1}$, $\omega^{\perp}_{to,2} = 125\ \text{cm}^{-1}$, $\omega^{\perp}_{p,2} = 55\ \text{cm}^{-1}$, $\omega^{z}_{to,1} = 135\ \text{cm}^{-1}$, $\omega^{z}_{p,1} = 283\ \text{cm}^{-1}$, $\omega^{z}_{to,2} = 154\ \text{cm}^{-1}$, $\omega^{z}_{p,2} = 156\ \text{cm}^{-1}$, $\gamma^{\alpha}_{j} = 3.5\ \text{cm}^{-1}$. The real parts of the dielectric functions $\varepsilon^{\perp}(\omega)$ and $\varepsilon^{z}(\omega)$ of Bi_2Se_3 are shown in Figure 2. There are two shaded regions, including regions A ($\omega^{\perp}_{to,1} < \omega < \omega^{z}_{to,1}$) where Bi_2Se_3 is type II HM ($\text{Re}(\varepsilon^z) > 0$, $\text{Re}(\varepsilon^{\perp}) < 0$) and region B ($\omega^{z}_{to,2} < \omega < 163\ \text{cm}^{-1}$) where Bi_2Se_3 is type I HM ($\text{Re}(\varepsilon^z) < 0$, $\text{Re}(\varepsilon^{\perp}) > 0$). Therefore, HPPs can be excited in these hyperbolic regions [56]. Although as a typical topological insulator, Bi_2Se_3 possesses unusual electron surface states, we firstly only consider its bulk optical response. When the frequencies are smaller than the bulk gap of 0.3 eV of Bi_2Se_3, the electronic contribution to permittivities (appearing in Equation (5) via ε^{z}_{∞}) is purely real. In addition, it is assumed that the valence bulk band is completely filled and the conduction band is empty, there are no free carriers present in the bulk. The doping electron surface states of Bi_2Se_3 is determined by its chemical potential μ_B that is located inside the bulk band gap. The electron surface states can be excluded by setting $\mu_B = 0.0$ eV.

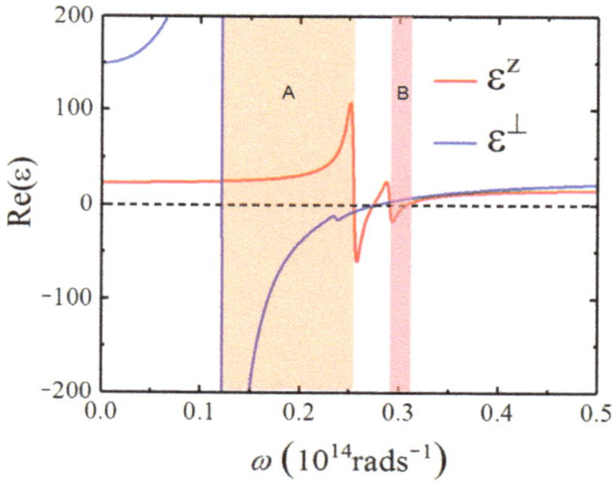

Figure 2. The real part of the dielectric function of Bi_2Se_3. The shaded regions denote the hyperbolic regions. Regions A and B denote type II ($\text{Re}(\varepsilon^{\perp}) \langle 0$, $\text{Re}(\varepsilon^z) \rangle 0$) and type I ($\text{Re}(\varepsilon^{\perp}) > 0$, $\text{Re}(\varepsilon^z) < 0$) hyperbolic bands, respectively.

3. Casimir Friction between Graphene-Covered Bi_2Se_3

In Figure 3, we show the dependences of Casimir friction on distance d for different configurations. As a natural hyperbolic material, hBN can also support HPPs which are located in the infrared regions. The permittivities of hBN can be found in Ref. [58]. Here we also show the Casimir friction between hBN bulk for comparison. The results of hBN, Bi_2Se_3, graphene, and graphene-covered Bi_2Se_3 are denoted in green dotted, red dashed, blue dot-dashed, and black solid lines, respectively. The chemical potential of graphene is fixed at $\mu = 0.2$ eV. It is seen that the Casimir friction coefficients of different configurations decrease as the distance increases. The friction coefficients of graphene and Bi_2Se_3 decrease slowly as the distance increases, while the friction coefficients of hBN and graphene-covered Bi_2Se_3 decrease rapidly as the distance increases. Although the friction coefficient of Bi_2Se_3 is smaller than that of hBN when the distance is less than 25 nm, it has an obvious enhancement when covered with graphene. Compared with that of Bi_2Se_3, the friction coefficient between graphene-covered Bi_2Se_3 can increase more than one order of magnitude when the chemical potential of graphene is 0.2 eV.

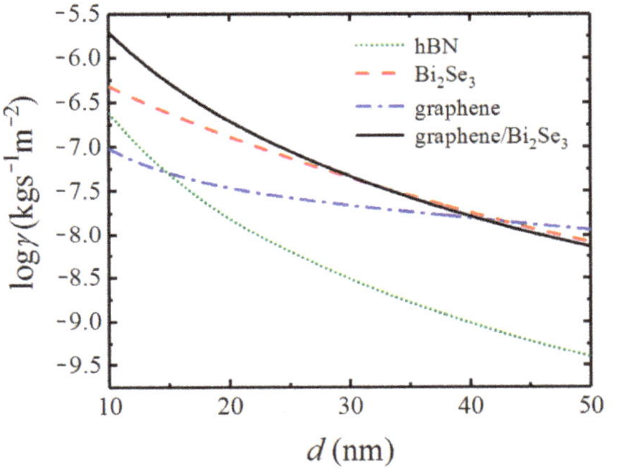

Figure 3. The dependence of Casimir friction coefficients for different configurations on separation distance d. The chemical potential of graphene is $\mu = 0.2$ eV. The base of the logarithm is 10 in this and the following figures.

Because Casimir friction is mainly from the exchange of the evanescent waves in the near-field region, we display the FSDs for different configurations to realize the contribution of different evanescent waves to Casimir friction in Figure 4. The results of hBN, Bi_2Se_3, graphene, and graphene-covered Bi_2Se_3 are displayed in green dotted, red dashed, blue dot-dashed, and black solid lines, respectively. The chemical potential of graphene is set to be $\mu = 0.2$ eV and the separation distance is $d = 20$ nm. It can be clearly seen from Figure 4 that there are two peaks corresponding to its hyperbolic bands located in the infrared frequency region for the FSD of hBN. Similarly, there are two peaks in the low-frequency regions for Bi_2Se_3. FSD has a strong peak corresponding to the resonance frequency of the HPPs on the surface of Bi_2Se_3 material, so the main contribution to quantum friction in Bi_2Se_3 comes from the HPPs. The FSD spectrum of graphene covers a very wide frequency range but with a relatively low value, which agrees that graphene can support p-polarized SPs with a wide frequency range. For graphene-covered undoped Bi_2Se_3, the HPPs supported by Bi_2Se_3 can be coupled with SPs supported by graphene, which results in the shift of the hybrid SPPPs toward higher frequencies. Therefore, the Casimir friction between graphene-covered Bi_2Se_3 is mainly from the SPPPs. We can qualitatively obtain the Casimir friction by judging the area covered by the curves of different configurations.

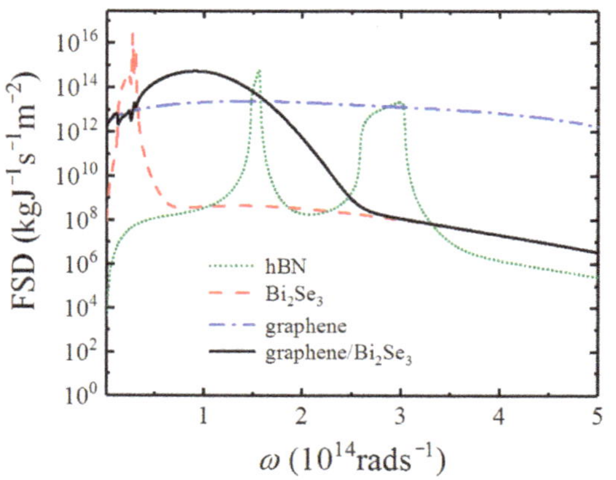

Figure 4. The force spectral densities for different configurations. The parameters used here are the same as those used in Figure 3.

To further analyze the physical mechanism of Casimir frictions for different configurations, we show the photon exchange functions of these configurations in Figure 5. The bright regions in the dark background denote the different surface polaritons supported by different materials. In Figure 5a, the bright regions are from the contributions of type I and type II hyperbolic bands of hBN, respectively. For the case of Bi_2Se_3 shown in Figure 5b. we can see that the hyperbolic bands also exist in the THz frequency range. For the photon exchange function between graphene sheets shown in Figure 5c, it covers a very wide frequency range since graphene can support SPs from THz to infrared frequencies. When Bi_2Se_3 is covered with graphene, the HPPs supported by Bi_2Se_3 can couple with SPs supported by graphene to form the SPPPs which is out of the hyperbolic band. The SPPPs move toward higher frequencies as the wave vector increases. Therefore, we can obtain that SPPPs dominate the Casimir friction between graphene-covered undoped Bi_2Se_3.

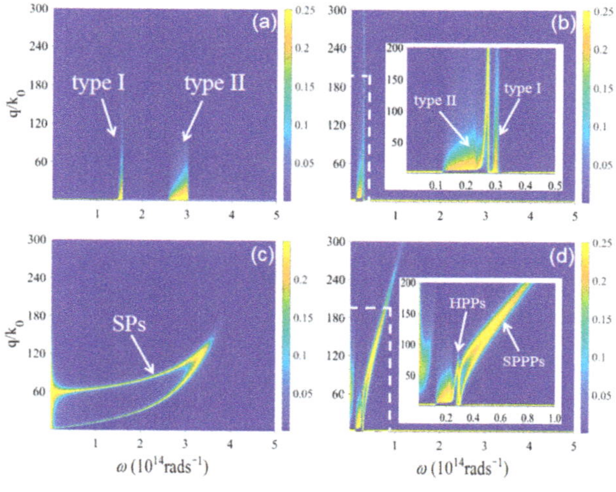

Figure 5. The photon exchange functions in different configurations. (**a**): hBN; (**b**): Bi_2Se_3; (**c**): graphene; (**d**): graphene-covered Bi_2Se_3. The chemical potential of graphene is $\mu = 0.2$ eV and the separation distance is $d = 20$ nm. The insets in (**b**,**d**) are the enlarged areas denoted by white dashed lines.

Because the characteristic of SPs excited by graphene as well as the coupled SPPPs is dependent on the chemical potential of graphene, we investigate the effect of chemical potential on the Casimir friction. In Figure 6, we display the friction coefficient on the separation distance d for different chemical potentials of graphene. The friction coefficient between graphene-covered Bi_2Se_3 with chemical potentials $\mu = 0.1$, $\mu = 0.2$, and $\mu = 0.3$ eV are plotted in black solid, blue dashed, and red dotted lines, respectively. We can see that the friction coefficients varying with distance exhibit different behavior for different chemical potentials. As the distance increases, the difference of Casimir friction between systems with different chemical potentials becomes smaller, which means that the contribution from the evanescent SPPPs becomes weaker as the distance increases.

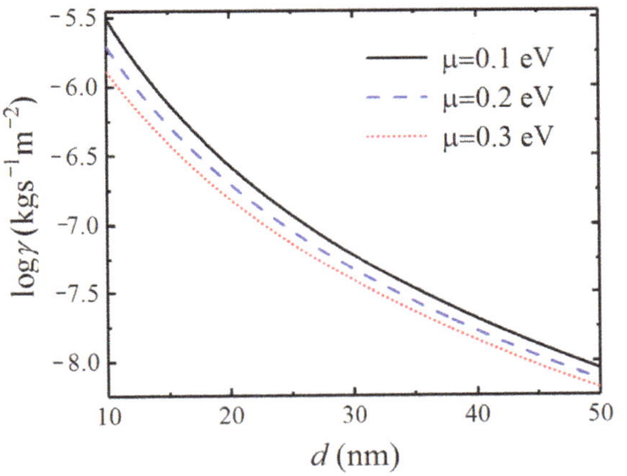

Figure 6. Dependence of the Casimir friction coefficient between graphene-covered Bi_2Se_3 on the separation distance for different chemical potentials of graphene.

In Figure 7, we show the dependence of the friction coefficient on the chemical potential between graphene-covered Bi_2Se_3 bulks. The friction coefficients with separation distances of 10, 20, and 30 nm are plotted in black solid, blue dashed, and red dotted lines, respectively. In Figure 7, we can see that the friction coefficients increase first and then decrease after reaching the maximum values as the chemical potential increases. In particular, the difference between the maximum value and minimum value of friction coefficients can reach about one order of magnitude for different chemical potentials for the separation distance being 10 nm. When the separation distance increases, the maximum and minimum values of Casimir friction still have an obvious difference in our considered chemical potential range. Therefore, the maximum value of the friction coefficient can be obtained by tuning the chemical potential of graphene.

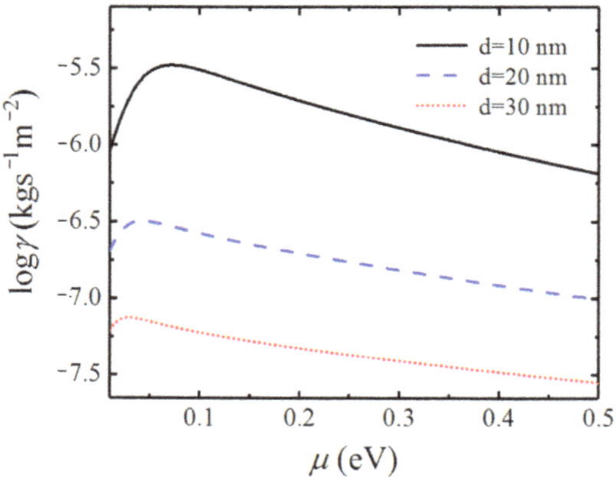

Figure 7. Dependence of the Casimir friction coefficient on chemical potential for different separation distances *d*.

In Figure 8, we show the FSD in order to understand the physical mechanism of the relationship between the friction coefficient and chemical potential. The separation distance between the top and bottom graphene-covered Bi_2Se_3 is fixed at $d = 20$ nm in the calculations. When the chemical potential of graphene is smaller ($\mu = 0.05$ eV), we can see that Casimir friction is mainly determined by the HPPs as shown by the solid black line. In this case, the coupling between HPPs and graphene plasmons is weak. The HPPs and graphene plasmon contribute to the Casimir friction separately. As the chemical potential increases ($\mu = 0.1$ eV), the coupling between HPPs and SPs occurs, resulting in the enhancement of SPPPs in a wide frequency region compared with the SPs supported by graphene as shown in red dashed lines in Figure 8. However, when the chemical potential continues to increase, the peak of SPPPs shifts toward higher frequencies with a decrease in magnitude. This will lead to the decrease in the Casimir friction. Therefore, we can demonstrate the coupling of graphene plasmon with HPPs supported by undoped Bi_2Se_3 can enhance the Casimir friction significantly between graphene-covered Bi_2Se_3. It is also possible to actively modulate the Casimir friction between such graphene/Bi_2Se_3 composite structures by controlling the gate voltage.

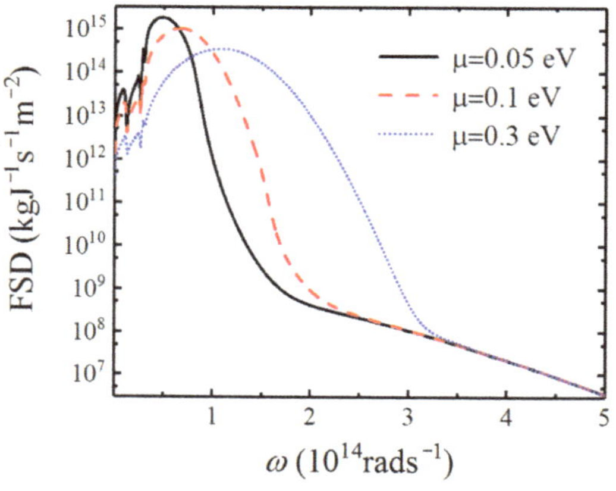

Figure 8. Force spectral densities of graphene-covered Bi_2Se_3 as a function of angular frequency for different chemical potentials at a fixed distance $d = 20$ nm.

4. Casimir Friction between Doped Bi_2Se_3 Bulks

We have investigated the Casimir friction between graphene-covered undoped Bi_2Se_3 in the previous sections. Then, we continue to study the Casimir friction between doped Bi_2Se_3 without covering graphene. Such a configuration can be realized by moving graphene away in Figure 1. In this configuration, the Bi_2Se_3 bulk behaves like an anisotropic material, while the two layers denoted in orange represent the top and bottom surface states. The electron surface states which behave as massless Dirac fermions exist when Bi_2Se_3 is doped; it plays a similar role as graphene in the graphene-covered undoped Bi_2Se_3 configuration. The fundamental density response functions of the surface states are the sheet conductivity σ_B and polarizability P, which have the following relation [56]

$$\sigma_B(q,\omega) = \frac{i\omega}{q^2}e^2P(q,\omega) \tag{6}$$

Under the condition of random-phase approximation for Dirac fermions, $P(q,\omega)$ can be obtained analytically [56]:

$$P(q,\omega) = -\frac{Nk_F}{2\pi\hbar v_F} - \frac{iN}{16\pi\hbar v_F}\frac{q^2}{\sqrt{q^2-k_\omega^2}}[G(\frac{k_\omega+2k_F}{q}) - G(\frac{k_\omega-2k_F}{q}) - i\pi] \tag{7}$$

where function $G(x)$ are expressed as $G(x) = ix\sqrt{1-x^2} - i\arccos x$. $N = 1$ is the number of Dirac cones. $k_F = |\mu_B|/\hbar v_F$ is the Fermi momentum with $v_F = \sqrt{\mu_{dc}\mu_B\tau_B/e}$ being the Fermi velocity, and μ_{dc} is the electron mobility. $k_\omega = (\omega + i\gamma_e)/v_F$ and $\gamma_e > 0$ is a phenomenological parameter that stands for electron scattering rate. After substituting the conductivity in Equation (4) with σ_B in Equation (6), we can calculate the Casimir friction between doped Bi_2Se_3 by employing Equation (1).

In Figure 9, we show the dependence of the Casimir friction coefficient on the separation distance d for different chemical potentials μ_B of doped Bi_2Se_3, which is similar to that shown in Figure 6. We can see that the Casimir friction coefficient decreases rapidly with the increase in separation distance d. When the separation distance is less than 20 nm, the friction coefficient is not sensitive to the chemical potential. When the separation distance becomes larger, the friction coefficient decreases as the chemical potential increases. Such phenomenon is different from that between graphene-covered undoped Bi_2Se_3 which is shown in Figure 6. In addition, comparing Figure 9 with Figure 3, we can also see that the friction coefficient between doped Bi_2Se_3 is larger than that between undoped Bi_2Se_3 when the separation distance is less than 20 nm. However, when the separation is larger than 20 nm, the difference in Casimir friction between undoped Bi_2Se_3 and doped Bi_2Se_3 is very small.

In Figure 10, we show the dependence of the Casimir friction coefficient on the chemical potential of doped Bi_2Se_3 for different separation distances d. When the separation distance is 10 nm, the friction coefficient almost does not vary with the chemical potential. That means the chemical potential of doped Bi_2Se_3 has little impact on the surface states that dominate the Casimir friction at such a small distance. The Casimir friction, in this case, is even larger than that between graphene-covered Bi_2Se_3 when the chemical potential is larger than 0.1 eV, which can be seen in Figure 7. For the cases with separation distances are 20 and 30 nm, the friction coefficients first increase and then decrease after reaching the maximum values as the chemical potential μ_B increase. However, the friction coefficients between doped Bi_2Se_3 are not as sensitive to the chemical potential as those between graphene-covered undoped Bi_2Se_3.

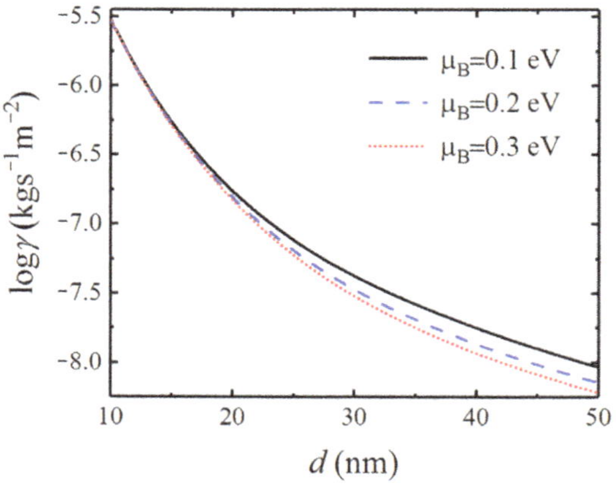

Figure 9. Dependence of the Casimir friction coefficient on the separation distances for different chemical potentials of Bi_2Se_3.

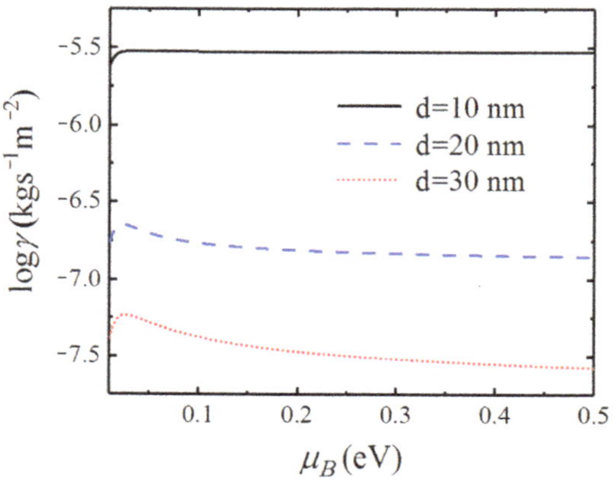

Figure 10. Dependence of the Casimir friction coefficient on chemical potential for different separation distances *d*.

5. Conclusions

In summary, we have studied the Casimir friction between graphene-covered Bi_2Se_3. Owing to the coupling of the HPPs supported by Bi_2Se_3 to the SPs excited by graphene, the Casimir friction between the proposed structures shows a significant enhancement compared to those between undoped Bi_2Se_3 materials. The Casimir friction can be modulated actively by tuning the chemical potential of graphene. It is found that the maximum value of Casimir friction can be obtained by choosing the appropriate chemical potential. Furthermore, Casimir friction between doped Bi_2Se_3 is also studied. When the chemical potential is larger than 0.1 eV, Casimir friction between doped Bi_2Se_3 is even larger than that between graphene-covered undoped Bi_2Se_3. Although more and more schemes have been proposed to enhance the Casimir force, it is still challenging to observe such a small force exactly in an experiment. Because it may play an important role in the nanoelectromechanical systems (NEMS), it is feasible to observe the Casimir force in the NEMS successfully. The results of this study are of great help to extend the research range of Casimir frictions and are meaningful to understanding the application of ultra-sensitive force detection of topological insulators.

Author Contributions: The manuscript was written through contributions of all authors. Conceptualization, T.Y. (Ting Yu) and R.L.; methodology, T.W. and T.Y. (Tianbao Yu); data curation, T.Y. (Ting Yu); validation, D.Z. and W.L.; writing—original draft preparation, T.Y. (Ting Yu); writing—review and editing, T.W.; supervision, Q.L.; funding acquisition, T.W. All authors have read and agreed to the published version of the manuscript.

Funding: This work is supported by the National Natural Science Foundation of China (Grant Nos. 12164027, 11704175, 12064025).

Institutional Review Board Statement: Not applicable.

Informed Consent Statement: Not applicable.

Data Availability Statement: The data that support the findings of this study are available upon reasonable request from the corresponding author.

Conflicts of Interest: The authors declare no conflict of interest.

References

1. Dorofeyev, I.; Fuchs, H.; Wenning, G.; Gotsmann, B. Brownian Motion of Microscopic Solids under the Action of Fluctuating Electromagnetic Fields. *Phys. Rev. Lett.* **1999**, *83*, 2402. [CrossRef]
2. Volokitin, A.I.; Persson, B.N.J. Noncontact friction between nanostructures. *Phys. Rev. B* **2003**, *68*, 155420. [CrossRef]
3. She, J.H.; Balatsky, A.V. Non-contact Friction and Relaxational Dynamics of Surface Defects. *Phys. Rev. Lett.* **2012**, *108*, 136101. [CrossRef]
4. Jentschura, U.D.; Janke, M.; DeKieviet, M. Theory of noncontact friction for atom-surface interactions. *Phys. Rev. A* **2016**, *94*, 022510. [CrossRef]
5. Stipe, B.C.; Mamin, H.J.; Stowe, T.D.; Kenny, T.W.; Rugar, D. Noncontact Friction and Force Fluctuations between Closely Spaced Bodies. *Phys. Rev. Lett.* **2001**, *87*, 096801. [CrossRef] [PubMed]
6. Kuehn, S.; Loring, R.F.; Marohn, J.A. Dielectric Fluctuations and the Origins of Noncontact Friction. *Phys. Rev. Lett.* **2006**, *96*, 156103. [CrossRef]
7. Saitoh, K.; Hayashi, K.; Shibayama, Y.; Shirahama, K. Gigantic Maximum of Nanoscale Noncontact Friction. *Phys. Rev. Lett.* **2010**, *105*, 236103. [CrossRef]
8. Pendry, J. Shearing the vacuum—quantum friction. *J. Phys. Condens. Matter* **1997**, *9*, 10301–10320. [CrossRef]
9. Farías, M.B.; Fosco, C.D.; Lombardo, F.; Mazzitelli, F.; López, A.E.R. Functional approach to quantum friction: Effective action and dissipative force. *Phys. Rev. D* **2015**, *91*, 105020. [CrossRef]
10. Barton, G. On van der Waals friction between two atoms at nonzero temperature. *New J. Phys.* **2011**, *13*, 043023. [CrossRef]
11. Reiche, D.; Dalvit, D.A.R.; Busch, K.; Intravaia, F. Spatial dispersion in atom-surface quantum friction. *Phys. Rev. B* **2017**, *95*, 155448. [CrossRef]
12. Zhao, R.; Manjavacas, A.; de Abajo, F.J.G.; Pendry, J. Rotational Quantum Friction. *Phys. Rev. Lett.* **2012**, *109*, 123604. [CrossRef] [PubMed]
13. Manjavacas, A.; Rodríguez-Fortuño, F.J.; de Abajo, F.J.G.; Zayats, A.V. Lateral Casimir Force on a Rotating Particle near a Planar Surface. *Phys. Rev. Lett.* **2017**, *118*, 133605. [CrossRef]
14. Høye, J.S.; Brevik, I. Casimir friction between a magnetic and a dielectric material in the nonretarded limit. *Phys. Rev. A* **2019**, *99*, 042511. [CrossRef]
15. Volokitin, A.I.; Persson, B.N.J. Dissipative van der Waals interaction between a small particle and a metal surface. *Phys. Rev. B* **2002**, *65*, 115419. [CrossRef]
16. Volokitin, A.I.; Persson, B.N.J. Resonant Photon Tunneling Enhancement of the van der Waals Friction. *Phys. Rev. Lett.* **2003**, *91*, 106101. [CrossRef]
17. Volokitin, A.I.; Persson, B.N.J. Adsorbate-Induced Enhancement of Electrostatic Noncontact Friction. *Phys. Rev. Lett.* **2005**, *94*, 086104. [CrossRef]
18. Volokitin, A.; Persson, B. Near-field radiative heat transfer and noncontact friction. *Rev. Mod. Phys.* **2007**, *79*, 1291–1329. [CrossRef]
19. Volokitin, A.I.; Persson, B.N.J. Theory of the interaction forces and the radiative heat transfer between moving bodies. *Phys. Rev. B* **2008**, *78*, 155437. [CrossRef]
20. Volokitin, A.; Persson, B.N.J. Quantum Friction. *Phys. Rev. Lett.* **2011**, *106*, 094502. [CrossRef]
21. Despoja, V.; Echenique, P.M.; Šunjić, M. Nonlocal microscopic theory of quantum friction between parallel metallic slabs. *Phys. Rev. B* **2011**, *83*, 205424. [CrossRef]
22. Farias, M.B.; Kort-Kamp, W.J.M.; Dalvit, D.A.R. Quantum friction in two-dimensional topological materials. *Phys. Rev. B* **2018**, *97*, 161407. [CrossRef]
23. Novoselov, K.S.; Geim, A.K.; Morozov, S.V.; Jiang, D.; Zhang, Y.; Dubonos, S.V.; Grigorieva, I.V.; Firsov, A.A. Electric field effect in atomically thin carbon films. *Science* **2004**, *306*, 666–669. [CrossRef] [PubMed]

24. Dell'Anna, L.; Merano, M. Clausius-Mossotti Lorentz-Lorenz relations and retardation effects for two-dimensional crystals. *Phys. Rev. A* **2016**, *93*, 053808. [CrossRef]
25. Novoselov, K.S.; Geim, A.K.; Morozov, S.V.; Jiang, D.; Katsnelson, M.I.; Grigorieva, I.V.; Dubonos, S.V.; Firsov, A.A. Two-dimensional gas of massless Dirac fermions in graphene. *Nature* **2005**, *438*, 197–200. [CrossRef]
26. Zhang, Y.B.; Tan, Y.W.; Stormer, H.L.; Kim, P. Experimental observation of the quantum Hall effect and Berry's phase in graphene. *Nature* **2005**, *438*, 201. [CrossRef]
27. Geim, A.K.; Novoselov, K.S. The rise of graphene. In *Nanoscience and Technology*; World Scientific: Singapore, 2009; pp. 11–19.
28. Berger, C.; Song, Z.; Li, X.; Wu, X.; Brown, N.; Naud, C.; Mayou, D.; Li, T.; Hass, J.; Marchenkov, A.N.; et al. Electronic confinement and coherence in patterned epitaxial graphene. *Science* **2006**, *312*, 1191–1196. [CrossRef] [PubMed]
29. Freitag, M.; Low, T.; Xia, F.; Avouris, P. Photoconductivity of biased graphene. *Nat. Photon* **2013**, *7*, 53–59. [CrossRef]
30. Stauber, T.; Peres, N.M.R.; Geim, A.K. Optical conductivity of graphene in the visible region of the spectrum. *Phys. Rev. B* **2008**, *78*, 085432. [CrossRef]
31. Falkovsky, L.A. Optical properties of graphene. *J. Phys. Conf. Ser.* **2008**, *129*, 012004. [CrossRef]
32. Falkovsky, L.A.; Pershoguba, S. Optical far-infrared properties of a graphene monolayer and multilayer. *Phys. Rev. B* **2007**, *76*, 153410. [CrossRef]
33. Mikhailov, S.A.; Ziegler, K.G. New Electromagnetic Mode in Graphene. *Phys. Rev. Lett.* **2007**, *99*, 016803. [CrossRef] [PubMed]
34. Ju, L.; Geng, B.; Horng, J.; Girit, C.; Martin, M.; Hao, Z.; Bechtel, H.A.; Liang, X.; Zettl, A.; Shen, Y.-R.; et al. Graphene plasmonics for tunable terahertz metamaterials. *Nat. Nanotechnol.* **2011**, *6*, 630. [CrossRef]
35. Jablan, M.; Buljan, H.; Soljačić, M. Plasmonics in graphene at infrared frequencies. *Phys. Rev. B* **2009**, *80*, 245435. [CrossRef]
36. Wang, T.-B.; Liu, N.-H.; Liu, J.-T.; Yu, T.-B. Quantum friction controlled by plasmons between graphene sheets. *Eur. Phys. J. B* **2014**, *87*, 185. [CrossRef]
37. Zhang, C.J.; Zhou, T.; Du, X.P.; Wang, T.B.; Liu, N.H. Enhancement of quantum friction via coupling of surface phonon polariton and graphene plasmons. *Acta Phys. Sin.* **2016**, *65*, 236801. [CrossRef]
38. Farias, M.B.; Fosco, C.D.; Lombardo, F.C.; Mazzitelli, F. Quantum friction between graphene sheets. *Phys. Rev. D* **2017**, *95*, 065012. [CrossRef]
39. Despoja, V.; Echenique, P.M.; Šunjić, M. Quantum friction between oscillating crystal slabs: Graphene monolayers on dielectric substrates. *Phys. Rev. B* **2018**, *98*, 125405. [CrossRef]
40. Luo, R.; Yang, J.-R.; Wang, T.-B.; Zhang, D.-J.; Liu, W.-X.; Yu, T.-B.; Liao, Q.-H. Enhancement of Casimir friction between graphene-covered hyperbolic materials. *Phys. Lett. A* **2021**, *387*, 127006. [CrossRef]
41. Fu, L.; Kane, C.L. Superconducting Proximity Effect and Majorana Fermions at the Surface of a Topological Insulator. *Phys. Rev. Lett.* **2008**, *100*, 096407. [CrossRef]
42. Qi, X.-L.; Zhang, S.-C. The quantum spin Hall effect and topological insulators. *Phys. Today* **2010**, *63*, 33–38. [CrossRef]
43. Hasan, M.Z.; Kane, C.L. Topological insulators. *Rev. Mod. Phys.* **2010**, *82*, 3045. [CrossRef]
44. Qi, X.L.; Zhang, S.C. Topological insulators and superconductors. *Rev. Mod. Phys.* **2011**, *83*, 1057. [CrossRef]
45. Karch, A. Surface plasmon and topological insultors. *Phys. Rev. B* **2011**, *83*, 245432. [CrossRef]
46. Fu, L.; Kane, C.L. Topological insulators with inversion symmetry. *Phys. Rev. B* **2007**, *76*, 045302. [CrossRef]
47. Hsieh, D.; Qian, D.; Wray, L.; Xia, Y.; Hor, Y.S.; Cava, R.J.; Hasan, M.Z. A topological Dirac insulator in a quantum spin Hall phase. *Nature* **2008**, *452*, 970–974. [CrossRef] [PubMed]
48. Zhang, H.; Liu, C.-X.; Qi, X.-L.; Dai, X.; Fang, Z.; Zhang, S.-C. Topological insulators in Bi2Se3, Bi2Te3 and Sb2Te3 with a single Dirac cone on the surface. *Nat. Phys.* **2009**, *5*, 438–442. [CrossRef]
49. Xia, Y.; Qian, D.; Hsieh, D.; Wray, L.A.; Pal, A.; Lin, H.; Bansil, A.; Grauer, D.; Hor, Y.S.; Cava, R.J.; et al. Observation of a large-gap topological-insulator class with a single Dirac cone on the surface. *Nat. Phys.* **2009**, *5*, 398–402. [CrossRef]
50. Hong, S.S.; Kundhikanjana, W.; Cha, J.J.; Lai, K.J.; Kong, D.S.; Meister, S.; Kelly, M.A.; Shen, Z.X.; Cui, Y. Ultrathin Topological Insulator Bi_2Se_3 Nanoribbons Exfoliated by Atomic Force Microscopy. *Nano Lett.* **2010**, *10*, 3118. [CrossRef]
51. Kong, D.; Dang, W.; Cha, J.J.; Li, H.; Meister, S.; Peng, H.; Liu, Z.; Cui, Y. Few-Layer Nanoplates of Bi_2Se_3 and Bi_2Te_3 with Highly Tunable Chemical Potential. *Nano Lett.* **2010**, *10*, 2245–2250. [CrossRef]
52. LaForge, A.D.; Frenzel, A.; Pursley, B.; Lin, T.; Liu, X.; Shi, J.; Basov, D.N. Optical characterization of Bi_2Se_3 in a magnetic field: Infrared evidence for magnetoelectric coupling in a topological insulator material. *Phys. Rev. B* **2010**, *81*, 125120. [CrossRef]
53. Akrap, A.; Tran, M.; Ubaldini, A.; Teyssier, J.; Giannini, E.; Van Der Marel, D.; Lerch, P.; Homes, C.C. Optical properties of Bi2Te2Se at ambient and high pressures. *Phys. Rev. B* **2012**, *86*, 235207. [CrossRef]
54. Wei, C.; Ren, S.F. Phonons of single quintuple Bi_2Te_3 and Bi_2Se_3 films and bulk materials. *Phys. Rev. B* **2011**, *83*, 094301. [CrossRef]
55. Zhou, L.; Carbotte, J.P. Hexagonal warping on spin texture, Hall conductivity and circular dichroism of topological Insulators. *Phys. Rev. B* **2014**, *89*, 165420. [CrossRef]
56. Wu, J.-S.; Basov, D.N.; Fogler, M. Topological insulators are tunable waveguides for hyperbolic polaritons. *Phys. Rev. B* **2015**, *92*, 205430. [CrossRef]
57. Ilic, O.; Jablan, M.; Joannopoulos, J.D.; Celanovic, I.; Buljan, H.; Soljačić, M. Near-field thermal radiation transfer controlled by plasmons in graphene. *Phys. Rev. B* **2012**, *85*, 155422. [CrossRef]
58. Zhao, B.; Guizal, B.; Zhang, Z.M.; Fan, S.; Antezza, M. Near-field heat transfer between graphene/hBN multilayers. *Phys. Rev. B* **2017**, *95*, 245437. [CrossRef]

59. Richter, W.; Becker, C.R. A Raman and far-infrared investigation of phonons in the rhombohedral V_2–VI_3 compounds Bi_2Te_3, Bi_2Se_3, Sb_2Te_3 and $Bi_2(Te_{1-x}Se_x)_3$ ($0 < x < 1$), $(Bi_{1-y}Sb_y)_2Te_3$ ($0 < y < 1$). *Phys. Status Solidi* **1977**, *84*, 619–628. [CrossRef]
60. Di Pietro, P.; Vitucci, F.M.; Nicoletti, D.; Baldassarre, L.; Calvani, P.; Cava, R.; Hor, Y.S.; Schade, U.; Lupi, S. Optical conductivity of bismuth-based topological insulators. *Phys. Rev. B* **2012**, *86*, 045439. [CrossRef]

nanomaterials

MDPI

Article

Magnetic Field Sensing Based on Whispering Gallery Mode with Nanostructured Magnetic Fluid-Infiltrated Photonic Crystal Fiber

Chencheng Zhang [1], Shengli Pu [1,2,*], Zijian Hao [1], Boyu Wang [1], Min Yuan [1] and Yuxiu Zhang [1]

[1] College of Science, University of Shanghai for Science and Technology, Shanghai 200093, China; 211560115@st.usst.edu.cn (C.Z.); 201840111@st.usst.edu.cn (Z.H.); boyuwang2002@163.com (B.W.); ym19821237760@163.com (M.Y.); 192282050@st.usst.edu.cn (Y.Z.)

[2] Shanghai Key Laboratory of Modern Optical System, University of Shanghai for Science and Technology, Shanghai 200093, China

* Correspondence: shlpu@usst.edu.cn; Tel.: +86-21-5527-1663

Abstract: A kind of novel and compact magnetic field sensor has been proposed and investigated experimentally. The proposed sensor consists of a tapered single mode fiber coupled with a nanostructured magnetic fluid-infiltrated photonic crystal fiber, which is easy to be fabricated. The response of magnetic fluid to magnetic field is used to measure the intensity of magnetic field via whispering gallery mode. The magnetic field-dependent shift in resonance wavelength is observed. The maximum magnetic field intensity sensitivity is 53 pm/mT. The sensor sensitivity is inversely proportional to the thickness of the photonic crystal fiber cladding.

Keywords: whispering gallery mode; microcavity resonator; magnetic field sensor; magnetic fluid; fiber sensor

1. Introduction

Compared with the traditional measurement methods, optical fiber magnetic field sensors have attracted great attention due to their outstanding virtues such as intrinsic safety, compact size, easy fabrication and immunity to electromagnetic interference. They have been widely applied in various fields such as electric power transmission, life science, medical health, geophysical research and military installations [1,2]. In principle, optical fiber is not sensitive to magnetic fields. Therefore, most of the proposed optical fiber magnetic field sensors realize magnetic field measurement by combining optical fiber with magnetic field sensitive materials [3].

Particularly, magnetic fluid (MF) is a kind of special nanostructured magneto-optic material, which consists of surfactant-coated magnetic nanoparticles and carrier liquid. It has unique optical properties such as dichroism, birefringence, Faraday effect and tunable refractive index under a magnetic field [4–10]. Furthermore, MF can be easily combined with optical fiber due to its fluidity. By infiltrating MF into the inner holes of microstructured fiber or coating the surface of specialty fiber with MF, many fiber-optic magnetic field sensors have been proposed based on various sensing technologies, such as multimode interference, Fabry–Perot interference, grating optics, Sagnac interference and surface plasma resonance [11–14]. The intensity of a magnetic field is usually measured by monitoring the wavelength shift of valley/peak in the output spectrum. Although these sensors have shown excellent sensing performance, the output spectral visibility is usually low and the full width at half maximum (FWHM) of the valley/peak is usually wide, which results in limited measurement accuracy and a low measurement resolution of magnetic field.

In recent years, the optical whispering gallery mode (WGM) resonator has shown enormous potential in sensing applications, such as biology [15–17], temperature [18–20], gas [21], humidity [22], laser [23], and magnetic field sensing [24–34]. Compared with the traditional optical resonators, the WGM resonator possesses high quality factor (Q), small mode volume and long lifetime of quantum photons. It is easy to achieve high-intensity output light with the WGM microcavity. Generally, the fabrication of WGM microcavity is based on mechanical deformation, while most of the mechanical processing would make fiber more fragile. Thus, the WGM tuning approach requires a delicate experimental apparatus. The tuning resolution is also rather limited. Those structures and methods may add the complexity of fabrication and increase the processing cost.

In this work, we propose and experimentally demonstrate a WGM sensor based on the photonic crystal fiber (PCF) infiltrated with MF. The MF serves as the magnetic field sensitive nanostructured material. The dependence of WGM resonance wavelength on magnetic field intensity is experimentally investigated. Furthermore, experimental results indicate that the influence of PCF diameter on the WGM resonant wavelength cannot be neglected. The proposed MF-infiltrated PCF WGM device possesses the features of high coupling efficiency and easy fabrication, which makes it a good candidate for application in magnetic field sensing and tunable optical filtering.

2. Fabrication and Experimental Details

Figure 1a shows the schematic diagram of the PCF coupled with a microfiber, where the PCF provides the cylindrical microcavity. The microfiber is obtained by tapering a single mode fiber (SMF) with the heat-and-pull method [35]. The final diameter of the tapering region is 2 μm. Figure 1b shows the WGM field distribution in x-y plane. The WGMs are effectively excited by evanescent coupling between the microfiber and the PCF microcavity. The WGMs circulate along the inner equator of the microcavity.

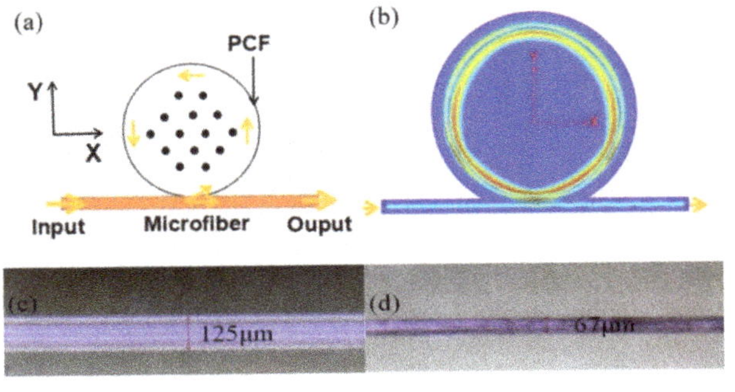

Figure 1. (**a**) Schematic diagram of the PCF infiltrated with nanostructured magnetic fluid coupled with a tapered microfiber. (**b**) Mode field distributions in x-y plane. (**c**) Micrograph of PCF before being infiltrated. (**d**) Micrograph of PCF after being infiltrated and corroded.

Figure 1c shows the micrograph of unfilled PCF (core/cladding diameter is 7/125 μm). There are periodic arrangement of air holes existing in the cross-section of PCF (hole diameter is 3 μm). Infiltration of MF in the 1.5 cm-long section of PCF is carried out by dropping one end of the PCF into MF. The MF infiltrates the PCF air holes under the influence of capillary force. The fabrication process is carried out at ambient temperature and pressure condition. The employed MF is water-based MF with surfactant-coated 10-nm-diameter Fe_3O_4 nanoparticles (provided by Hangzhou Jikang New Materials Co., LTD, Hangzhou, China). The initial volume fraction is 33%. The refractive index (RI) of MF applied in this work is 1.36 and its volume fraction is around 9.7% (diluted with distilled water and measured by a refractometer (A670, Hanon, Jinan, China)). In theory, MF with higher concentration (viz. higher RI) may be greatly sensitive to the magnetic field. Then, two ends of the MF-infiltrated PCF are sealed with UV glue. The sealed PCF is put into the hydrofluoric (HF) acid solution with a concentration of 20%. The HF acid can corrode the outer wall of the sealed PCF. By controlling the corrosion time, cylindrical microcavity with uniform diameter could be obtained. Figure 1d shows the micrograph of the PCF after being infiltrated with MF and corroded by HF acid. The obtained PCF diameter (d) is about 67 μm. Experiments indicate that the HF begin permeating into the air holes and, hence, breaks the PCF microcavity when the corrosion time is longer than 70 min.

Figure 2 shows the schematic diagram of the experimental setup for investigating magnetic field sensing properties. An electromagnet and a highly stable light source with emitting light, covering wavelength range of 1450–1650 nm, were employed. The output power of the light source is 12 mW. The MF-infiltrated PCF WGM structure was placed in the middle of the electromagnet and carefully aligned to make the PCF parallel to the magnetic field. The magnetic field strength was adjusted by changing the magnitude of the supply current. It is measured simultaneously by a gauss meter with an accuracy of 0.1 mT. The transmission spectrum is monitored by the optical spectrum analyzer (OSA, Yokogawa AQ6370C, Tokyo, Japan). The dips near 1530 nm wavelength are selected for monitoring and analyzing (see the violet rectangle in Figure 3 below).

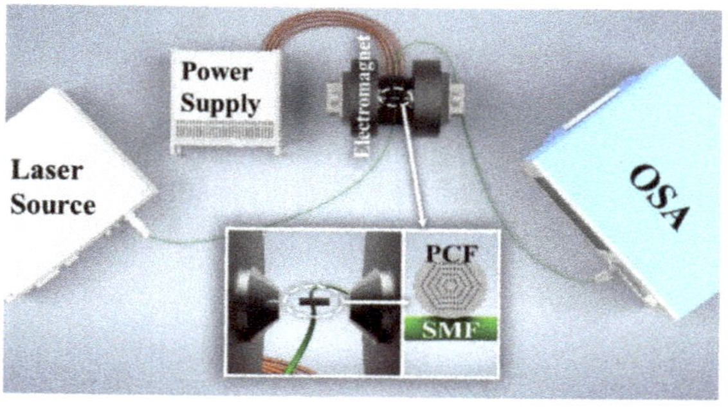

Figure 2. Schematic of experimental setup for investigating the sensing properties. SMF: single mode fiber, PCF: photonic crystal fiber, OSA: optical spectrum analyzer.

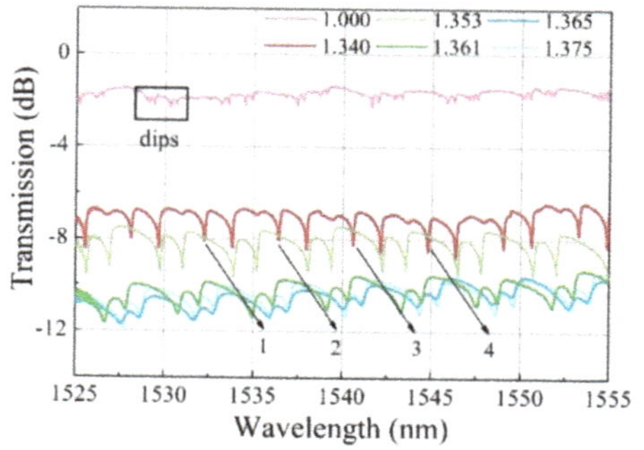

Figure 3. Transmission spectra of the sensing structure when the coupling region is immersed in glycerin solution of different refractive indices.

The coupling distance between PCF and microfiber is accurately controlled by two three-axis translation stages. The detailed process is as follows: first, one end of the SMF is connected with the light source and the other end is connected with the OSA. Then, two ends of the SMF tapering region are fixed on the glass slide with UV glue and the tapering region is partially suspended. Next, the PCF microcavity is fixed on the three-dimensional moving platform. Through adjusting the translation stage, the PCF microcavity can be moved slowly close to the tapering region for accurate coupling, which is synchronously observed by a CCD camera connected with the computer.

3. Sensing Principle and Experimental Results

The operating principle of the sensing structure is based on monitoring the shift of resonant wavelength. After the incident light enters the tapering region of the SMF, WGMs can be excited. WGM resonant wavelength should satisfy the following resonance condition [36].

$$m\lambda = 2\pi n_{eff} R \tag{1}$$

where m is an integer, λ is the resonant wavelength, n_{eff} refers to effective refractive index (ERI) and R represents the outer radius of the PCF microcavity. Under the condition of fixed microcavity radius, resonant wavelength λ is affected by ERI n_{eff}. The relationship between λ and n_{eff} is given as

$$\Delta\lambda = \left(\frac{\lambda}{n_{eff}}\right)\Delta n_{eff} \tag{2}$$

Equation (2) indicates that the resonant wavelength will shift towards the long wavelength with the increase in ERI.

To verify the above theory, the influence of ERI on resonant wavelength has been experimentally investigated and the results are shown in Figure 3. The sensing device consists of a tapered microfiber and a MF-infiltrated PCF microcavity (1.5 cm long, 125 μm in diameter) was placed on the experimental platform. The ambient RI near the coupling region was changed by dropping glycerin solution with different concentrations. When the ambient RI increased from 1.00 to 1.375, the resonant wavelength shifted towards the long wavelength direction (see the black arrows labeled as 1, 2, 3, 4 in Figure 3). The experimental results are in good agreement with the theory, which indicates that the sensor is sensitive to the change in the ambient RI.

For magnetic field measurement, MF is employed as the magnetic field sensitive material. MF has both the magnetic properties of solid magnetic materials and the fluidity of liquid materials. Under zero magnetic field, the magnetic nanoparticles within MF are dispersed randomly and uniformly in carrier liquid, as shown in Figure 4a, which are assigned to the influence of surfactant and Brownian motion of magnetic nanoparticles. When the magnetic field is applied, the magnetic nanoparticles within MF are rapidly magnetized. Due to the dipolar interaction of inter-nanoparticles, the magnetic nanoparticles favor agglomeration along the magnetic field direction and form nanochain-like structures [37–39]. Figure 4b schematically shows the chains along the field direction. Figure 4c shows the magnetic nanoparticle chain distribution in PCF under applied magnetic field. Then, the dielectric constant and hence the RI of MF will change with magnetic field. The relationship between RI of MF and magnetic field is given by the following Langevin-like function [40]

$$n_{MF}(H) = \begin{cases} n_0, & H < H_{c,n} \\ (n_S - n_0)\left[\coth\left(\alpha\frac{H - H_{c,n}}{T}\right) - \frac{T}{\alpha(H - H_{c,n})}\right] + n_0, & H > H_{c,n} \end{cases} \tag{3}$$

where n_0 is the RI of MF at zero magnetic field. n_s is the saturated value of MF's RI. H is the magnetic field strength in Gs. $H_{c,n}$ is the critical value of magnetic field strength. T is the thermodynamic temperature in Kelvin. α is the adjusting parameter. Thus, the resonant wavelength will shift with the magnetic field, which is fundamental for magnetic field sensing.

Figures 5 and 6 show the transmission spectra of the as-fabricated sensing structures under different magnetic field intensities. The diameters of the PCF microcavity are 67 and 75 μm, respectively. When the magnetic field intensity changes from 0 to 1.8 mT, the resonant wavelength changes greatly, as shown in Figures 5a and 6a. Then, as the magnetic field intensity further increases to 14 mT, the resonant wavelength redshifts further as shown in Figures 5b and 6b, but the shift speed decreases remarkably.

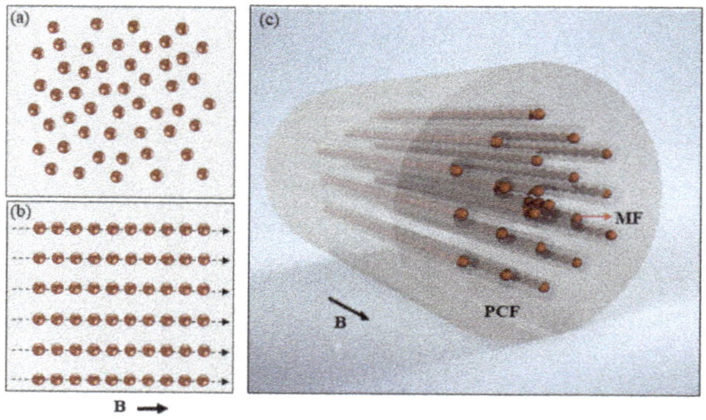

Figure 4. (**a**) Schematic of magnetic nanoparticles within magnetic fluid under zero magnetic field. (**b**) Schematic of magnetic nanoparticle columns within magnetic fluid under applied magnetic field. (**c**) Magnetic nanoparticle column distribution in PCF.

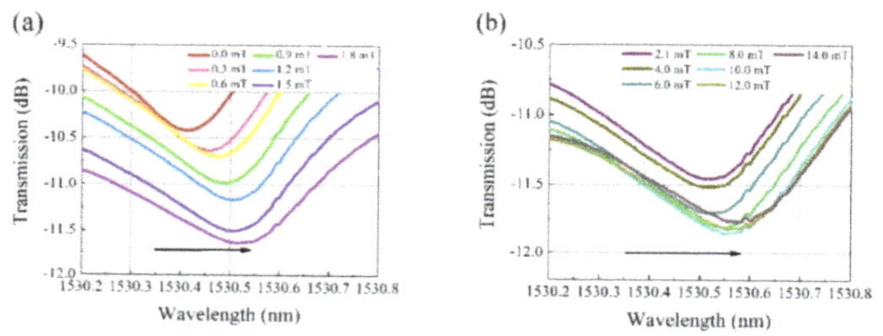

Figure 5. Transmission spectra at different magnetic field intensities for the PCF microcavity with diameter of 67 μm. The magnetic field strength is in the range of (**a**) 0 to 1.8 mT and (**b**) 2.1 to 14 mT.

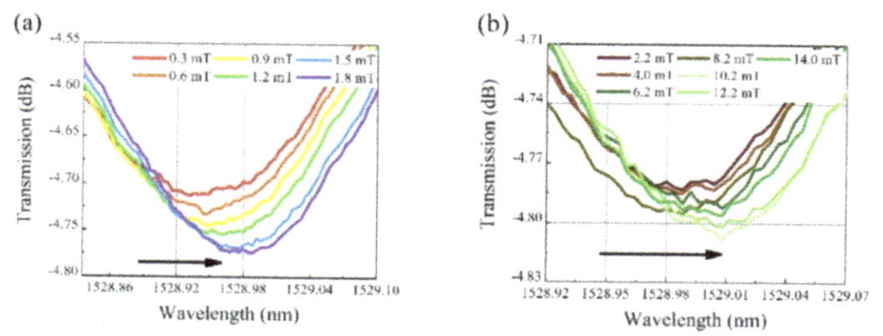

Figure 6. Transmission spectra at different magnetic field intensities for the PCF microcavity with diameter of 75 μm. The magnetic field strength is in the range of (**a**) 0.3 to 1.8 mT and (**b**) 2.2 to 14 mT.

Figure 7 explicitly shows the corresponding resonant wavelength as a function of magnetic field intensity. Figure 7 indicates that, in the magnetic field range from 0 to 1.8 mT, the linear responsivities of the two sensing structures are 53 pm/mT and 26 pm/mT, respectively. However, in the magnetic field range from 2 to 14 mT, their linear responsivities are 6 pm/mT and 2 pm/mT, respectively. The maximum sensitivity is obtained to be 53 pm/mT. Higher sensitivity is obtained at weak magnetic field. At high magnetic field, the sensitivity decreases, which may be assigned to the tendency to saturation state of MF. In addition, the sensitivity is inversely proportional to PCF microcavity diameter. The smaller the PCF microcavity diameter is, the higher the sensitivity will be. Therefore, the proposed WGM sensors with higher sensitivity can be obtained by reducing the cladding thickness of the PCF microcavity.

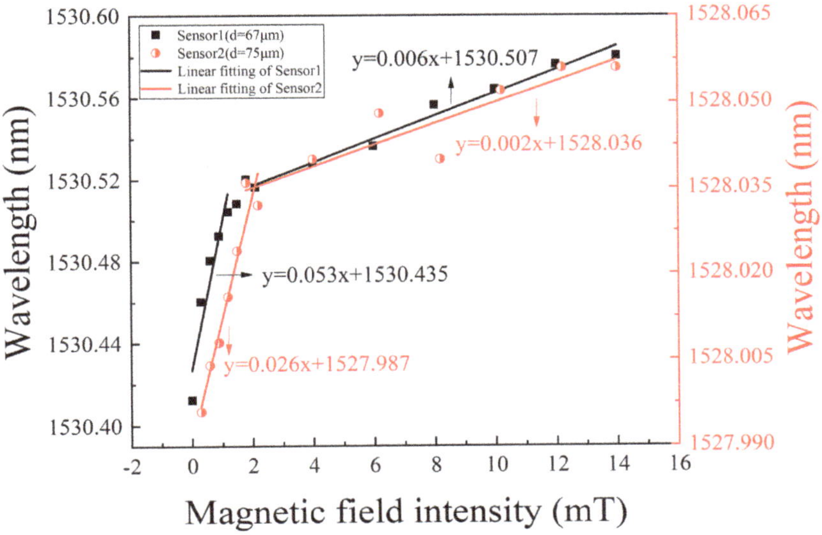

Figure 7. WGM resonant wavelength as a function of applied magnetic field for the PCF microcavity with diameter of d = 67 and 75 μm, respectively.

For comparison, Table 1 lists the sensing structures, fabrication methods and sensing performance of the related optical fiber WGM magnetic field sensors. The mainly adopted structures and fabrication methods include pressure-assisted arc discharging on silica capillary or hollow fiber to obtain microbottles, or adjusting related fusion parameters on SMF to fabricate microspheres. It is obvious from Table 1 that the sensitivity of the proposed PCF WGM sensor is better than those of the structures based on hollow microbubble, borosilicate glass, microsphere and hollow microbottle, but lower than those of the structures based on silica microcapillary, fuse-discharging silica capillary and side-polished SiO_2 microsphere. However, compared with the above-mentioned structures and fabrication methods, the PCF microcavity employed in this work has natural cylindrical microcavity and the fabrication process is simple and easy to repeat. Thus, the MF-infiltrated PCF WGM sensor is promising for highly sensitive weak magnetic field measurement in the future.

Table 1. Sensing performance of various optical fiber WGM magnetic field sensors.

Sensing Structure	Fabrication Method	Maximum Sensitivity	Ref.
Hollow microbubble	Fuse-and-blow technique	0.081 pm/mT	[24]
Borosilicate glass	HF etching + silicon nitride ceramic heating + glassblowing method	4.0 pm/mT	[25]
Microsphere	HF etching + fusing + electrode discharging method	5.036 pm/mT	[26]
Silica microbubble	Fusing + discharging method	15.1 pm/mT	[27]
Hollow microbottle	Pressure-assisted air discharging + HF etching method	25.21 pm/mT	[28]
PCF	Tapering method	−61.86 pm/mT	[29]

Table 1. *Cont.*

Sensing Structure	Fabrication Method	Maximum Sensitivity	Ref.
Silica microcapillary	Hydrogen flame heating + air-pumped swelling method	75.7 pm/mT	[30]
Silica capillary	Fuse-discharging + increasing gas pressure + heated and soften method	8.45 pm/Gs (84.5 pm/mT)	[31]
SiO_2 microsphere	Side-polishing method	0.0108 nm/Oe (108 pm/mT)	[32]
PCF	HF etching method	53 pm/mT	This work

4. Conclusions

In summary, a kind of magnetic field sensor based on WGM excited in a MF-infiltrated PCF resonator has been demonstrated. The maximum magnetic field intensity sensitivity is obtained to be 53 pm/mT. Experimental results indicate that the cladding thickness of PCF microcavity influences the sensitivity of the sensor. The proposed device possesses the desirable features of high coupling efficiency and easy fabrication, which make it promising for application in magnetic field sensing and tunable optical filtering. The PCF-based WGM resonance structure is also simple and compact.

Author Contributions: Conceptualization, S.P. and C.Z.; software, Z.H.; validation, C.Z. and B.W.; data curation, C.Z., M.Y. and Y.Z.; writing—original draft preparation, C.Z.; writing—review and editing, S.P.; project administration, S.P.; funding acquisition, S.P. All authors have read and agreed to the published version of the manuscript.

Funding: This research is supported by National Natural Science Foundation of China (Grant No. 62075130, 61675132), Shanghai "Shuguang Program" (Grant No. 16SG40), and Shanghai Talent Development Fund (Grant No. 201529).

Data Availability Statement: The original contributions presented in the study are included in the article. Further inquiries can be directed to the corresponding author.

Conflicts of Interest: The authors declare no conflict of interest.

References

1. Zhao, Y.; Tong, R.-J.; Xia, F.; Peng, Y. Current status of optical fiber biosensor based on surface plasmon resonance. *Biosens. Bioelectron.* **2019**, *142*, 111505. [CrossRef] [PubMed]
2. Lenz, J.; Edelstein, S. Magnetic sensors and their applications. *IEEE Sens. J.* **2006**, *6*, 631–649. [CrossRef]
3. Zhao, Y.; Liu, X.; Lv, R.-Q.; Zhang, Y.-N.; Wang, Q. Review on optical fiber sensors based on the refractive index tunability of ferrofluid. *J. Light. Technol.* **2017**, *35*, 3406–3412. [CrossRef]
4. Janssen, J.; Perenboom, J. Magneto-optical phenomena in magnetic fluids: The influence of orientation of anisotropic scatterers. *J. Magn. Magn. Mater.* **1989**, *81*, 14–24. [CrossRef]
5. Hasmonay, E.; Dubois, E.; Bacri, J.-C.; Perzynski, R.; Raikher, Y.L.; Stepanov, V.I. Static magneto-optical birefringence of size-sorted γ-Fe_2O_3 nanoparticles. *Eur. Phys. J. B* **1998**, *5*, 859–867. [CrossRef]
6. Taketomi, S. Magnetic fluid's anomalous pseudo-cotton mouton effects about 107 times larger than that of nitrobenzene. *Jpn. J. Appl. Phys.* **1983**, *22*, 1137–1143. [CrossRef]
7. Xu, M.; Ridler, P.J. Linear dichroism and birefringence effects in magnetic fluids. *J. Appl. Phys.* **1997**, *82*, 326–332. [CrossRef]
8. Horng, H.; Hong, C.-Y.; Yang, S.Y.; Yang, H.C. Novel properties and applications in magnetic fluids. *J. Phys. Chem. Solids* **2001**, *62*, 1749–1764. [CrossRef]
9. Gao, R.; Lu, D.-F.; Zhang, Q.; Xin, X.-J.; Tian, Q.-H.; Tian, F.; Wang, Y.-J. Temperature compensated three-dimension fiber optic vector magnetic field sensor based on an elliptical core micro fiber Bragg grating. *Opt. Express* **2020**, *28*, 7721–7733. [CrossRef] [PubMed]
10. Yuan, M.; Pu, S.L.; Li, D.H.; Li, Y.X.; Hao, Z.J.; Zhang, Y.X.; Zhang, C.C.; Yan, S.K. Extremely high sensitivity magnetic field sensing based on birefringence-induced dispersion turning point characteristics of microfiber coupler. *Results Phys.* **2021**, *29*, 104743. [CrossRef]
11. Li, X.-G.; Zhao, Y.; Cai, L.; Zhou, X. Measurement of magnetic field and temperature based on fiber-optic composite interferometer. *IEEE Trans. Instrum. Meas.* **2017**, *66*, 1906–1911. [CrossRef]

12. Zhou, X.; Li, X.G.; Li, S.G.; An, G.-W.; Cheng, T.L. Magnetic field sensing based on SPR optical fiber sensor interacting with magnetic fluid. *IEEE Trans. Instrum. Meas.* **2019**, *68*, 234–239. [CrossRef]
13. Zhao, Y.; Wu, D.; Lv, R.-Q.; Li, J. Magnetic field measurement based on the Sagnac interferometer with a ferrofluid-filled high-birefringence photonic crystal fiber. *IEEE Trans. Instrum. Meas.* **2016**, *65*, 1503–1507. [CrossRef]
14. Sun, B.; Fang, F.; Zhang, Z.; Xu, J.; Zhang, L. High-sensitivity and low-temperature magnetic field sensor based on tapered two-mode fiber interference. *Opt. Lett.* **2018**, *43*, 1311–1314. [CrossRef] [PubMed]
15. Hao, Y.; Guo, Z. Integrated sensor with a whispering-gallery mode and surface plasmonic resonance for the enhanced detection of viruses. *J. Opt. Soc. Am. B* **2021**, *38*, 2855–2862. [CrossRef]
16. Guo, Z.; Qin, Y.; Chen, P.; Hu, J.; Zhou, Y.; Zhao, X.; Liu, Z.; Fei, Y.; Jiang, X.; Wu, X. Hyperboloid-drum microdisk laser biosensors for ultrasensitive detection of human IgG. *Small* **2020**, *16*, 2000239. [CrossRef]
17. Swaim, J.D.; Knittel, J.; Bowen, W.P. Detection limits in whispering gallery biosensors with plasmonic enhancement. *Appl. Phys. Lett.* **2011**, *99*, 243109. [CrossRef]
18. Liao, J.; Yang, L. Optical whispering-gallery mode barcodes for high-precision and wide-range temperature measurements. *Light. Sci. Appl.* **2021**, *10*, 32. [CrossRef] [PubMed]
19. Wang, E.; Cheng, P.; Li, J.; Cheng, Q.; Zhou, X.; Jiang, H. High-sensitivity temperature and magnetic sensor based on magnetic fluid and liquid ethanol filled micro-structured optical fiber. *Opt. Fiber Technol.* **2020**, *55*, 102161. [CrossRef]
20. Wang, Z.; Mallik, A.K.; Wei, F.; Wang, Z.; Rout, A.; Wu, Q.; Semenova, Y. Thermo-optic tuning of a nematic liquid crystal-filled capillary whispering gallery mode resonator. *Opt. Express* **2021**, *29*, 23569–23581. [CrossRef]
21. Li, H.Y.; Sun, B.; Yuan, Y.G.; Yang, J. Guanidine derivative polymer coated microbubble resonator for high sensitivity detection of CO_2 gas concentration. *Opt. Express* **2019**, *27*, 1991–2000. [CrossRef] [PubMed]
22. Reinis, P.K.; Milgrave, L.; Draguns, K.; Brice, I.; Alnis, J.; Atvars, A. High-sensitivity whispering gallery mode humidity sensor based on glycerol microdroplet volumetric expansion. *Sensors* **2021**, *21*, 1746. [CrossRef]
23. Yan, J.; Wang, D.N.; Li, X. An erbium-doped whispering-gallery-mode microlaser for sensing. *J. Light. Technol.* **2021**, *39*, 5177–5182. [CrossRef]
24. Guo, Y.; Zhang, Y.D.; Sun, H.Y.; Zhu, F.X.; Yi, G.; Wang, J.F. Magnetic-field tuning whispering gallery mode based on hollow microbubble resonator with Terfenol-D-fixed. *Appl. Opt.* **2019**, *58*, 8889–8893. [CrossRef]
25. Freeman, E.; Wang, C.-Y.; Sumaria, V.; Schiff, S.J.; Liu, Z.W.; Tadigadapa, S. Chip-scale high Q-factor glassblown microspherical shells for magnetic sensing. *AIP Adv.* **2018**, *8*, 065214. [CrossRef]
26. Xia, J.; Wang, F.Y.; Xiong, S.D.; Zhou, G.Y.; Hu, Y.M.; Yao, Q. Novel magnetometer basedon magnetic fluid coated whispering gallery ode resonator. In Proceedings of the 2019 18th International Conference on Optical Communications and Networks, Huangshan, China, 5–8 August 2019; pp. 1–3.
27. Liu, Y.L.; Zhang, H.; Fan, M.S.; Liang, C.Z.; Lei, J. Bidirectional tuning of whispering gallery modes in a silica microbubble infiltrated with magnetic fluids. *Appl. Opt.* **2020**, *59*, 1–8. [CrossRef]
28. Liu, W.; Li, W.Y.; Wang, R.; Xing, E.B.; Jing, N.; Zhou, Y.R.; Tang, J.; Liu, J. Magnetic sensor based on WGM hollow microbubble resonator filled with magnetic fluid. *Opt. Commun.* **2021**, *497*, 127148. [CrossRef]
29. Mahmood, A.; Kavungal, V.; Ahmed, S.S.; Kopcansky, P.; Zavisova, V.; Farrell, G.; Semenova, Y. Magnetic field sensing using whispering-gallery modes in a cylindrical microresonator infiltrated with ferronematic liquid crystal. *Opt. Express* **2017**, *25*, 12195–12202. [CrossRef]
30. Zhu, S.; Shi, L.; Liu, N.Y.; Xu, X.B.; Zhang, X.L. Magnetic field sensing using magnetic-fluid-filled optofluidic ring resonator. *Microfluid. Nanofluid.* **2017**, *21*, 156. [CrossRef]
31. Hou, F.Y.; Zhang, X.B.; Wang, Z.J.; Yang, L.; Sun, W.; Yang, Y.; Dong, Y.H.; Huang, Y.; Wang, T.Y. Magnetic fluid infiltrated microbottle resonator sensor with axial confined mode. *IEEE Photon. J.* **2020**, *12*, 1–9. [CrossRef]
32. Huang, H.M.; Yu, Y.; Zhou, L.J.; Tao, Y.Y.; Yang, J.B.; Zhang, Z.R. Whispering gallery modes in a microsphere attached to a side-polished fiber and their application for magnetic field sensing. *Opt. Commun.* **2020**, *478*, 126366. [CrossRef]
33. Li, Y.X.; Pu, S.L.; Zhao, Y.L.; Zhang, R.; Jia, Z.X.; Yao, J.L. All-fiber-optic vector magnetic field sensor based on side-polished fiber and magnetic fluid. *Opt. Express* **2019**, *27*, 35182–35188. [CrossRef] [PubMed]
34. Zhang, Y.-N.; Zhu, N.; Gao, P.; Zhao, Y. Magnetic field sensor based on ring WGM resonator infiltrated with magnetic fluid. *J. Magn. Magn. Mater.* **2020**, *493*, 165701. [CrossRef]
35. Vega, F.; Torres, C.; Mattos, L. Method of producing tapered fibers. In Proceedings of the 22nd Congress of the International Commission for Optics: Light for the Development of the World, Puebla, Mexico, 15–19 August 2011.
36. Suter, J.D.; Fan, X. Overview of the optofluidic ring resonator: A versatile platform for label-free biological and chemical sensing. In Proceedings of the Annual International Conference of the IEEE Engineering in Medicine and Biology Society (EMBC), Minneapolis, MN, USA, 3–6 September 2009; pp. 1042–1044. [CrossRef]
37. Yin, J.; Yan, P.; Chen, H.; Yu, L.; Jiang, J.; Zhang, M.; Ruan, S. All-fiber-optic vector magnetometer based on anisotropic magnetism-manipulation of ferromagnetism nanoparticles. *Appl. Phys. Lett.* **2017**, *110*, 231104. [CrossRef]
38. Bertoni, G.; Torre, B.; Falqui, A.; Fragouli, D.; Athanassiou, A.; Cingolani, R. Nanochains formation of superparamagnetic nanoparticles. *J. Phys. Chem. C* **2011**, *115*, 7249–7254. [CrossRef]

39. Zhang, Z.; Guo, T.; Zhang, X.-J.; Xu, J.; Xie, W.-P.; Nie, M.; Wu, Q.; Guan, B.-O.; Albert, J. Plasmonic fiber-optic vector magnetometer. *Appl. Phys. Lett.* **2016**, *108*, 101105. [CrossRef]
40. Hong, C.Y.; Horng, H.E.; Yang, S.Y. Tunable refractive index of magnetic fluids and its applications. *Phys. Status Solid. C* **2004**, *1*, 1604–1609. [CrossRef]

nanomaterials

MDPI

Article

Superfluorescence of Sub-Band States in C-Plane $In_{0.1}Ga_{0.9}N/GaN$ Multiple-QWs

Cairong Ding *, Zesheng Lv, Xueran Zeng and Baijun Zhang

State Key Laboratory of Optoelectronic Materials and Technologies, School of Electronics and Information Technology (School of Microelectronics), Sun Yat-sen University, Guangzhou 510275, China; lvzsh@mail2.sysu.edu.cn (Z.L.); zengxr@mail.sysu.edu.cn (X.Z.); zhbaij@mail.sysu.edu.cn (B.Z.)

* Correspondence: dingcr@mail.sysu.edu.cn; Tel.: +86-020-841-102-79

Abstract: Superfluorescence is a collective emission from quantum coherent emitters due to quantum fluctuations. This is characterized by the existence of the delay time (τ_D) for the emitters coupling and phase-synchronizing to each other spontaneously. Here we report the observation of superfluorescence in c-plane $In_{0.1}Ga_{0.9}N/GaN$ multiple-quantum wells by time-integrated and time-resolved photoluminescence spectroscopy under higher excitation fluences of the 267 nm laser and at room temperature, showing a characteristic τ_D from 79 ps to 62 ps and the ultrafast radiative decay (7.5 ps) after a burst of photons. Time-resolved traces present a small quantum oscillation from coupled $In_{0.1}Ga_{0.9}N/GaN$ multiple-quantum wells. The superfluorescence is attributed to the radiative recombination of coherent emitters distributing on strongly localized subband states, $E_{e1}{\rightarrow}E_{hh1}$ or $E_{e1}{\rightarrow}E_{lh1}$ in 3nm width multiple-quantum wells. Our work paves the way for deepening the understanding of the emission mechanism in the $In_{0.1}Ga_{0.9}N/GaN$ quantum well at a higher injected carrier density.

Keywords: InGaN/GaN multiple-quantum wells; superfluorescence; time-resolved photoluminescence; photoluminescence; collective emissions

1. Introduction

Spontaneous emissions (SEs) of photons occur because of coupling between excited two-level systems and vacuum modes of the electromagnetic field, such as the process of fluorescence that is commonly used in displays and lighting—effectively stimulated by its zero-point fluctuations. C-plane InGaN/GaN quantum wells (QWs) as the prevailing active layer have been studied extensively due to their promising applications in group-III nitride semiconductor optoelectronic devices [1–4]. Commonly accepted explanations for emission features are the spatial localization of carriers due to random alloy fluctuations, indium compositional fluctuations and well width fluctuations [5–18], the quantum-confined Stark effect (QCSE) because it spatially separates electron and hole wave functions and reduces the wave function overlap in the QWs [19–25] and the screening of the QCSE under a high excitation that affects the excitation density-dependent emission energy of InGaN/GaN MQWs, for example, a very strong emission from quantum-dot-like states [26], high energy emission band [27–29] and stimulated emission on the high-energy side in thick QWs [30–32].

Rime-resolved photoluminescence (TRPL) presents fast and slow decay behaviors over several orders of magnitude due to weakly-localized carriers transferring among different energy levels easily [33,34]. The decay dynamics of carriers at a time scale longer than 200 ns is ascribed to agglomerated localization centers on nearly isolated islands [35].

Light–matter interactions can also create and manipulate collective many-body phases in solids. Lasing and cooperative emissions (superradiance SR and superfluorescence SF) form a family of collectively coherent emissions—a collective emission that results in a burst of photons with ultrafast radiative decay. SR was described by Dicke in 1954 [36] followed by the experimental observation in 1973 [37]. Several SF emissions were reported [38,39].

SR and SF are characterized by an accelerated radiative decay time $\tau \propto \tau_{SE}/N$, where the exponential decay time τ_{SE} of SE from the uncoupled two level systems is shortened by the number of coupled emitters N [40,41]. In addition, the peak intensity of SF and SR increases superlinearly with N, $I_{SF} \propto N^2$.

SF emissions are typically characterized by the unique existence of the delay time $\tau_D \propto \log(N)/N$ for the development of macroscopic spontaneous quantum coherence due to quantum fluctuations [36–38] and Burnham–Chiao ringing which can't be observed in all experiments primarily due to a spatial averaging effect [42] and pure superfluorescence [41].

In this case, we observe a previously unreported unique behavior of the SF associated with the uniformity of localized states in $In_{0.1}Ga_{0.9}N/GaN$ QWs especially at high excitation level with which generally, the manufactured devices like laser diodes are always operating [43]. Hence, it is necessary to deepen the understanding of the emission mechanism at high excitation levels.

2. Materials and Methods

InGaN/GaN MQWs schematized in Figure 1a were grown on (0001) c-plane sapphire substrates in a metal organic chemical vapor deposition (MOCVD) system with the Thomas Swan closely spaced showerhead reactor. The In content in the 3 nm thick InGaN QW is 10%. The measured period of the multiple QW system is about 13 nm, leading to 10 nm barrier width.

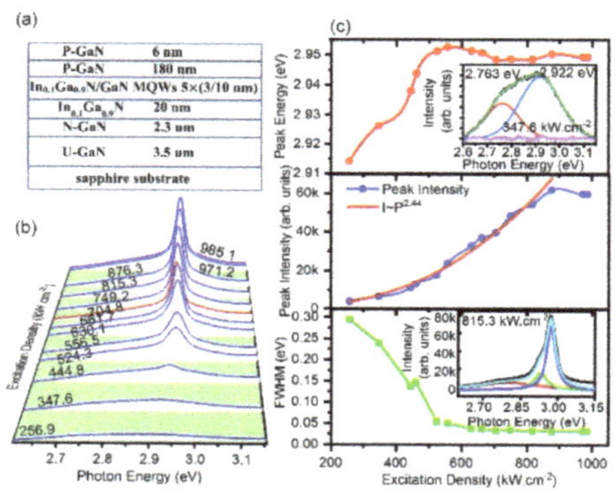

Figure 1. (Color Online) Excitation fluence dependent photoluminescence behaviors of $In_{0.1}Ga_{0.9}N/GaN$ MQWs (**a**) Schematic diagram of material structure. (**b**) Color Mapping of PL intensities and the emission photon energy at different excitation fluences (P) from 256.9 kW.cm^{-2} to 985.1 kW.cm^{-2}. (**c**) Top, the shift of the peak emission energy with the excitation fluence. The inset exhibits the flat and broad PL spectrum line at low excitation fluence; Middle, the peak emission intensity that increases superlinearly with the excitation fluence, corresponding to a power-law dependence with an exponent α = 2.4 (red line); Bottom, the narrowing of the FWHM with the excitation fluence obtained from the peak fitting to the spectra in (**b**). The inset shows several typical spectra. Different color lines in subfigures (**c**) present fitting peaks to spectra with a Lorentzian for various excitation power densities.

Time-integrated and time-resolved photoluminescence measurements were performed at room temperature using a regeneratively amplified femtosecond Ti: sapphire laser with a repetition rate of 1 kHz operating around 800 nm. The 25 fs pulses duration were tripled to provide a 267 nm excitation source with the maximum power of about 20 mW (Micro-5 + Legend Elite USP HE+, Coherent), corresponding to 2.686×10^9 photons per pulse.

The excitation laser output was reflected to a prism in order to obtain a 267 nm laser. The 267 nm laser beam was in alignment with the center height of the slit by way of two mirrors and apertures and focused by a singlet lens with the focal length of 70 mm onto the sample $In_{0.1}Ga_{0.9}N/GaN$ MQWs in the way of an oblique incidence. The diameter d of the laser spot is 1.488×10^{-4} cm, the area of the laser spot is 1.738×10^{-8} cm^2. We were thus able to reach the maximum excitation densities of 1150.8 kW cm^{-2}. The backward emission from the sample was collected perpendicular to the sample surface by a duplet lens system (f = 70 mm and 320 mm, d = 70 mm) and converged onto the input slit of a double-grating monochromator (SpectraPro hrs-300, PI) and detected by a CCD camera (Optronis SRU-ED, Kehl, Germany) which are connected to an ultraviolet-visible streak camera system with 2 ps time resolution and an instrument response function of 6 ps (Optoscope SC-10 SRU-ED, Optronis GmBH, Kehl, Germany). The 300 mm spectrograph has two gratings with 300 and 1200 lines per mm. The input slit of the spectrograph is a vertical slit; while the output slit of the streak camera is a horizontal one. We adjusted slit width to 10 μm. The time base is 0~288 ps under the fastest sweep speed of 15 ps/mm. The pump power (P) is controlled with a variable neutral density filter.

3. Results and Discussion

Figure 1 presents the excitation density-dependent photoluminescence (PL) spectra of $In_{0.1}Ga_{0.9}N$/GaN MQWs for the excitation fluence (P) from 256.9 to 985.1 kW cm^{-2}. The emission spectra were measured in the range of 2.6–3.1 eV. At the low excitation less than 347.6 kWcm^{-2} (Figure 1b), the emission exhibits a flat and broad PL spectrum line shape with full width at half maximum (FWHM) of about 277 meV (inset, Figure 1c top panel), which is the convolution of two emission peaks at the center energy 2.763 eV and 2.922 eV. These two emissions are attributed to sequential spontaneous emissions of shallow impurities [44]. However, at the excitation density of 444.8 kW cm^{-2}, a narrow emission shoulder peak of 2.945 eV appears at the high energy side in PL. As excitation fluence is increased, the emission peak energy is blue-shifted up to 2.960 eV at the exaction density of 661.7 kW cm^{-2}.

The quantitative analysis of spectrally integrated PL which is best fitted with a Lorentzian for various excitation power densities is shown in Figure 1c (the inset shows the fitting to the spectrum at the excitation fluence 815.3 kWcm^{-2}, bottom panel). As the excitation fluence is increased from the excitation density of 461.6 kW cm^{-2}, PL peak energies first present the blueshift feature (top panel). That will be discussed in detail in Figure 2. As the excitation fluence is further increased from 704.8 kW cm^{-2} to 985.1 kW.cm^{-2}, the emission peak energy is red-shifted from 2.960 eV to 2.949 eV presumably owing to renormalization of the emission energy from the dense coherent coupling. The peak intensity increases superlinearly over three orders of magnitude (middle panel) according to a power-law dependence with an exponent of $\alpha = 2.4 \pm 0.1$ deviating from the theoretically expected value of $\alpha = 2$ [42] presumably by assuming that more localized states devote to nonlinear emissions. At the excitation density exceeds 985.1 kW.cm^{-2}, the emission peak intensity reduces due to the horizontal transmission of the delocalized carriers [45] and renormalization of the emission energy. Delocalized carriers face a higher number of fast non-radiative recombination centers leading to an increase of non-radiative losses. The full-width at half-maximum (FWHM) of the PL spectrum is excitation sensitive and reduced to about 30 meV after the excitation threshold (bottom panel). The emission-line narrowing is one of the observed features of the SF [39]. Remarkably, narrow emission lines are similar to the 36 meV linewidth measured on a single GaN quantum dot at room temperature [46]. These narrow contributions are, therefore, fully compatible with the expected emission from a single localized state.

Figure 2 discussed the blueshift of the emission peak energy. Generally, the PL peak energy of the InGaN/GaN MQWs can be expressed as [47]:

$$E_p = E_g + \Delta E_{strain} - \Delta E_{QCSE} + \Delta E_{QSE} \quad (1)$$

where E_g is the bandgap of relaxed InGaN alloys, ΔE_{strain} represents the blueshift caused by the compressive strain, ΔE_{QCSE} represents the well width dependent red shift in the effective band gap due to the quantum confined Stack effect (QCSE) [48], and ΔE_{QSE} represents the shift caused by the intersubband transition due to the quantum size effect (QSE). In our InGaN MQWs, the first two items would not be changed by the excitation power. Thus, the ΔE_{QCSE} and ΔE_{QSE} could be responsible for the blue-shift of the PL spectrum as the excitation fluence increases. As we know, the QCSE usually causes red-shifts owing to the tilt of energy bands. Moreover, under a high excitation fluence, the QCSE may be weakened by the huge number of photogenerated carriers, resulting in a decrease of QCSE-caused red-shift, i.e., a PL peak blue-shift, However, this process does not result in a time-delayed blueshift emission peak, as observed in Figure 3d. Thus, the discussed blue-shift is more likely caused by ΔE_{QSE}, the changes in the intersubband transitions.

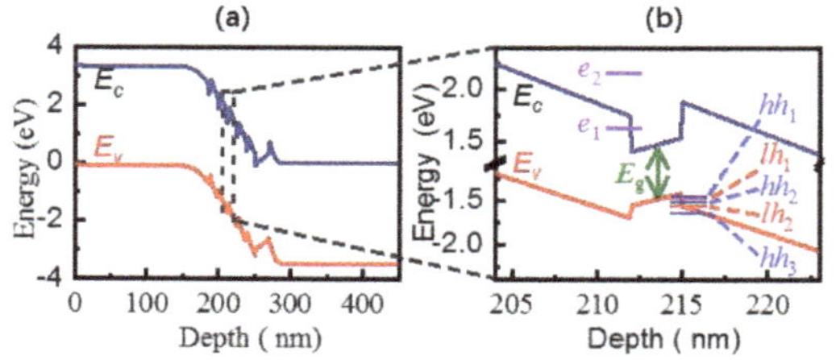

Figure 2. (Color Online) (**a**) Energy band diagram simulated by Silvaco TCAD software at 0 V bias; (**b**) energy band details of the InGaN QWs, together with the calculated subband energies.

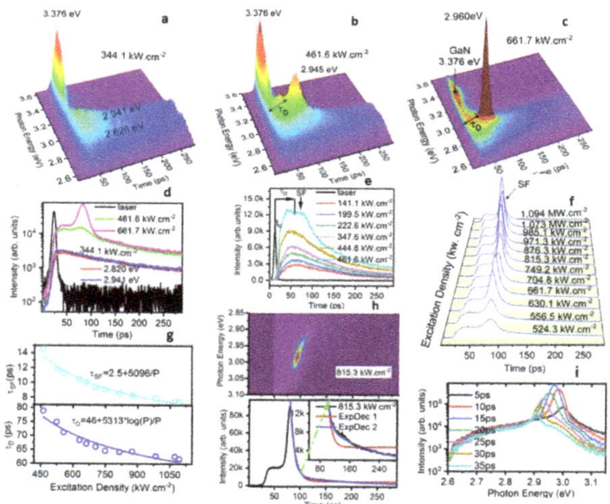

Figure 3. (Color Online) Steak camera images and time-resolved photoluminescence spectroscopy for different excitation fluences. (**a**) weak excitation 344.1 kW.cm^{-2}. (**b**) Observation of delayed bursts of radiation with threshold excitation 461.6 kW cm^{-2}. (**c**) Strong excitation 661.7 kW cm^{-2}. The delay time (τ_D) is defined (**d**) Time-resolved photoluminescence spectroscopy for three excitation fluences above. The laser given here shows the initial time of the emission. Obviously, the delay emission is observed after the threshold excitation. (**e**,**f**) Extracted time-resolved emission intensity traces for different excitation fluences over 100 meV energy window in detail. The initial time is indicated as t_0. τ_D, the delay time; SF, superfluorescence. (**g**) Upper, decay time (τ_{SF}) as a function of the excitation fluence (P), fitted using the superfluorescence model (cyan line): $\tau_{SF} = 2.5 + 5096/P$. (**h**) Upper, streak camera image obtained at an excitation fluence of 815.5 kW cm^{-2}. Lower, time-resolved emission intensity trace (with an enlarged view) presents the definite difference from traces fitted by ExpDec 1 (red line) and ExcDey 2 (blue line) functions. (**i**) The dynamic redshift of time-resolved slices integrated over a 5 ps time window at different time delays on a semi-log scale.

To clarify the mechanism in the observed cooperative emission and related peak shifts, the energy band diagram of the structure was simulated at 0 V, using Silvaco TCAD software, as shown in Figure 2a. Besides, a band detail of the QW is illustrated as Figure 3b in which the subband energies with different n were calculated using:

$$E_n = \frac{\hbar^2 \pi^2}{2m^* L_w^2} n^2 \tag{2}$$

where $\hbar$ is the reduced Planck constant, L_W is the well width, and effective masses of $In_{0.1}Ga_{0.9}N$ were taken in the calculations are: $m_e = 0.21\ m_0$ for electrons, $m_{lh} = 1.22\ m_0$ for light holes and $m_{hh} = 2.00\ m_0$ for heavy holes [49]. The results indicate that the subband energies in the conduction band are more discrete owing to the smaller effective mass of electrons. The energy difference between E_{e1} and E_{e2} is about 0.597 eV, far exceeding the measured PL peak variation of 15 meV. Thus, we deduce that the transitions from E_{e1} to E_{lhn} or E_{hhn} should be dominant in our experiments. In the case of 267 nm excitation ($hv = 4.662$ eV), photogenerated holes distribute the multiple subbands of the valence band, and then two processes will happen synchronously: (1) photogenerated holes at high energy levels will relax to the lowest E_{hh1} gradually; (2) radiative recombination occurs between photogenerated electrons at E_{e1} and photogenerated holes at multiple subband energies. As the excitation power increases, more and more holes will relax to the lowest E_{hh1} before they are recombined with electrons. Therefore, at the excitation fluence 461.6 kW cm^{-2}, the carriers' transition from E_{e1} to E_{hh1} enhances gradually, leading to the increasing PL intensity at 2.945 eV and a narrow FWHM. Meanwhile, the delay time is observed as the excitation density increases due to the coherent emitters (Figure 3b–f), also supporting the relaxation process of the holes. At slightly higher excitation fluence, the states at E_{hh1} will be completely filled by photogenerated and relaxed holes owing to their limited state density. Thus, the transition from E_{e1} to E_{hh1} tends to saturation, and the transition from E_{e1} to E_{lh1} dominates gradually, leading to the emission peak blueshift for the excitation fluence from 461.6 to 704.8 kW cm^{-2}. Furthermore, the delay time will decrease gradually with the excitation fluence (Figure 3g, lower panel) because coherent emitters from E_{e1}-E_{lh1} transitions form earlier spontaneously than coherent emitters from E_{e1}-E_{hh1} transitions. In addition, the calculated E_{lh1} is about 0.013 eV higher than E_{hh1}, which also agrees with the observed blue shift from 2.945 eV to 2.960 eV. Above all, the observed narrow emissions should come from the localized intersubband transitions of the InGaN wells, in which SF from E_{e1}-E_{hh1} transition is dominant at the excitation fluence of 461.6 kW cm^{-2} and SF from E_{e1}-E_{lh1} coherent transitions will be enhanced with an increasing excitation fluence. Thus, the blueshift of the emission from the QWs is confirmed to originate from the confined subband levels when the well width is smaller than 3–4 nm [44].

Figure 3 presents streak camera images and extracted time-resolved emission intensity traces (over 100 meV energy window) for various excitation densities in detail and quantitative analysis. At the low excitation fluence of 344.1 kW cm^{-2}, two emission peak energies at 2.820 and 2.941 eV exhibit almost the same dynamics for spontaneous emissions of shallow impurities (Figure 3a). The rise time of the emission peak energy at 2.941 eV (blue trace, Figure 3d) is 8ps shorter than that of the emission peak energy at 2.820 eV (red trace, Figure 3d), showing carriers competitively populate on shallow impurity energy levels. The initial time of these two emissions is almost the same as that of the emission of GaN top surface. With P = 461.6 kW cm^{-2}, the peak emission energy shows the blueshift to 2.945 eV and its dynamic process presents the distinct delay time τ_D before the rising edge and a more shortened radiative decay and long radiative decay (green trace in Figure 3b,d). As the excitation fluence of 661.7 kW cm^{-2}, in addition to the τ_D, we cannot simply describe the time-resolved PL as the bi-exponential decay (Figure 3c and purple trace, Figure 3d).

Figure 3e,f show extracted time-resolved emission traces for different excitation fluences over 100 meV energy window. In time-resolved PL decay measurements with low excitation fluence less than 444.8 kW cm^{-2} (yellow trace in Figure 3e), we did not observe a significant modification in the decay of the spontaneous emission, which can be described as a momo-exponential decay with 1/e decay time of τ_{SE} = 1180 ps.

At a slightly higher excitation fluence 461.6 kW cm^{-2} (cyan trace in Figure 3e), after the delay time τ_D, we observed a more pulse named SF with 1/e decay time of τ_{SF} = 67 ps in comparison to the PL decay of SE with 1/e decay times of τ_{SE} = 1180 ps. The delay time (τ_D) was calculated as the period between the start point of the emission and the rising edge of 2.945 eV. As the excitation densities P increase up to 524.3 kW cm^{-2} (Figure 3f), after the τ_D, a rising edge and the accelerated PL decay of the emission with a moderately long tail cannot be described by a mono-or bi-exponential function of the spontaneous emission, as shown in Figure 3c. As excitation fluences increase more and more, such as at 815.3 kW cm^{-2} (Figure 3h, lower panel), the time-resolved emission intensity trace displays a small oscillatory superfluorescence shape [41,42] during 118~165 ps, which is different from traces fitted by mono-exponential (red line) and bi-exponential(blue line) decay functions. However, a pure superfluorescence pulse has a single hyperbolic-secant shape [41]. Since the penetrating depth of laser has effects on the distribution of state densities of a high-gain medium, When the laser incidents on c-plane $In_{0.1}Ga_{0.9}N$/GaN multi-QWs, the sample may thus be considered as being divided into a number of concentric shells of decreasing density, the coupling of carriers in multi-QWs' layers by diffraction makes spatial averaging which washes out the ringing pulse period.

As excitation fluence increases further, the decay time (τ_{SF}) shortens to 8 ps (upper panel in Figure 3g), exceeding the instrument response limit by 6 ps at FWHM. The cyan line presents that τ_{SF} is fitted to the excitation fluence P for excitation densities over 461.6 kW cm^{-2}, $\tau_{SF} \propto 1180\ \alpha/P$, where α is a scaling factor. The initial number of coupled emitters N is proportional to the excitation fluence P/α. Here, a fixed value of τ_{SE} = 1180 ps was used, as found for spontaneous emission of $In_{0.1}Ga_{0.9}N$/GaN QWs at low excitation density. The proportional relation is in accord with $\tau_{SF} \propto \tau_{SE}/N$ of cooperative emissions [36,41]. The lower amplitudes of slower components are observed (Figure 3c). Consequently, the SF decay rate should converge towards the decay rate of spontaneous emission.

Furthermore, a shortening of the SF delay time (lower panel in Figure 3g), after which the photon burst is emitted, is observed. This characteristic of SF is a consequence of the time it takes for the individual dipoles to become phase-locked and scales with the number N of excited coupled emitters according to $\tau_D \propto \log(N)/N$ [38,39,41]. The data are fitted to $\tau_D = 46 + 5313 \times log(P)/P$ (blue line, Figure 3g lower panel) by assuming N to be proportional to the excitation fluence P. The drastic shortening of the radiative lifetime and the delay time attest to the observed emission being superfluorescence. The delay time τ_D decreases for high excitation fluences owing to the increased interaction among emitters, which is in accordance with the red-shift of the emission peak energy for higher excitation fluences discussed in Figure 1c, upper panel.

Figure 3h,i, present the decay image of the emission peak energy of 2.967 eV with the excitation density of 815.3 kW cm^{-2} and dynamic red-shift slices integrated over a 5 ps time window at different times delays in the semi-logarithmic scale. Although the observed image of SF is alike to the Fermi-edge SF in $In_{0.2}Ga_{0.8}As$/GaAs [50], as time goes on, the dynamic red-shift in our $In_{0.1}Ga_{0.9}N$/GaN is attributed to the coherently radiative emission of carriers distributed on E_{e1} and E_{hh1}, carriers reduce during the relaxation (Figure 2b). It takes much more time for photogenerated holes at high energy levels to relax to the lowest E_{hh1} gradually and form coherent emitters, resulting in a longer τ_D. SF coherent states result from these narrow localization states. In addition, a dynamical red-shift is owing to the renormalization of the emission energy from the coherent coupling. The SF mission maybe undergoes the evolution process of the coherent emission to the noncoherent emission due to the large lateral dimensions or number fluctuations within the coherent SF state.

Figure 4 presents streak camera images of SF emission for four typical excitation fluences and spectrally integrated time-resolved PL traces (over 100 meV energy window) for six typical excitation fluences from 317.9 to 871.2 kW cm^{-2} in a new batch of the sample. In Figure 4a, the build-up evolution of the SF photon burst is clearly shown on the base of the delay time τ_D. At the P = 473.3 kW cm^{-2}, SF appears clearly after the delay time. At the P = 871.2 kWcm^{-2}, a brighter image after the τ_D is shown due to photon bursts of SF.

Figure 4b presents normalized time-resolved PL traces for different fluences in order to observe the evolution of the τ_D and the decay trace of the emission more clearly. As the excitation fluence increase, $\tau_D = 26 + 4490 \times \log(P)/P$ (Figure 4b, inset), which confirms to $\tau_D \propto \log(N)/N$ of SF [38,39,41]. After photon bursts of SF, the decay trace of the emission shortens much with the increase of excitation fluences. In Figure 4c, after the overlap (the black arrow) of decay traces, the emission intensity increase a little bit (red traces, the magnified image in green circle of the inset) at the excitation fluence of 698.5 kW cm^{-2}, which means some weak oscillation is observed in the time domain compared with the decay trace under the weak excitation fluence of 317.9 kW cm^{-2} (black trace). The 3 nm QWs' width may contribute to the SF, Quantum confined localization states are coupled with QWs, which results in the oscillation. The oscillation is a fundamental property of SF [42]. Figure 4d shows the fitting to fluorescence traces of SE and SF at excitation fluences of 317.9 kW cm^{-2} and 769.9 kW cm^{-2} respectively. The small oscillation was considered to be ignored due to error problems during the fitting. Decay traces are simply described by a bi-exponential function, there are no damped oscillations involved. The rising edge of the SE is described as a biphasic dose response function given that one part of carriers distribute on upper levels and another part of photons may start to form cooperative mode because the end of the rising edge shows the characteristics of the formation process. The rising edge of SF was described as a revised Double Boltzmann function, or a polynomial function given that the SE and the formation of cooperative states occur in this period. The horizontal transmission was observed with the side geometry (not shown there). These observations were consistently reproducible. In addition, the output face of the sample was not limited to a small spatially-resolve region, thus, the remarkable oscillation was not demonstrated in our experiments at room temperature. The strong SF ring is observed in 1–10 µm lateral dimensions of a homogeneous system [38], or a small detect region [41,51] at low temperature. The output of the sample as a whole is the sum of the radiation from each of the small dimensions, and this "spatial averaging" washes out the ringing or oscillation.

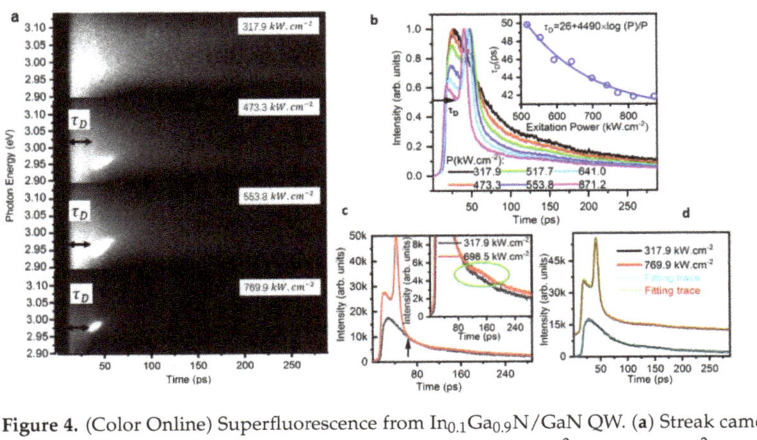

Figure 4. (Color Online) Superfluorescence from $In_{0.1}Ga_{0.9}N/GaN$ QW. (**a**) Streak camera images of SF emission for four typical excitation fluences, 317.9 kWcm^{-2}, 473.3 kWcm^{-2}, 553.8 kWcm^{-2}, 769.9 kW.cm^{-2}, the τ_D means the delay time. (**b**) Spectrally integrated time-resolved PL traces (over 100 meV energy window) for different excitation fluences (P units: kW.cm^{-2}). Inset, the excitation fluence of the delay time was best fitted with a function $\tau_D = 26 + 4490 \times \log(P)/P$. (**c**) Time-resolved PL traces for excitation fluences of 698.5 kW cm^{-2} (red) compared to the time-resolved PL trace at 317.9 kW cm^{-2} (black), the insets show a small ringing circled in the green ellipse, the black arrow points at the overlap of the falling edge of emission traces. (**d**) The fitting to fluorescence traces of SE and SF at excitation fluences of 317.9 kW.cm^{-2} and 769.9 kW.cm^{-2} respectively.

Given the previous observations, our experiments revealed that the SF from the highly excited $In_{0.1}Ga_{0.9}N/GaN$ QWs at room temperature can be confirmed unambiguously by the evidence for the characteristic delay time τ_D, $\tau_{SF} \propto \tau_{SE}/N$, emission line narrowing with the increasing of excitation fluences, and quadratic P dependence of the peak emission intensity I_{SF}. This is attributed to the coherent radiative emissions of carriers distributing on the subband E_{e1} and E_{hh1} or E_{lh1} of localized states in $In_{0.1}Ga_{0.9}N/GaN$ MQWs at higher injected carrier density. The oscillation of the emission pulse was not observed clearly owing to spatial averaging and room temperature. The results opened a door for us to research collectively emissive characteristics of quantum wells. The fitting to the SF pulse further demonstrated its characteristics.

As the SF critically depends on the low decoherence and low inhomogeneous energy variance, the long dephasing time (T_2^*) [52] which is mainly dependent on the stringent quality of the material structure, high oscillator strength, and small inhomogeneous broadening. In the time domain, T_2^* can be directly measured with free-induction decay, optical nutation, photon-echo, etc. SF occurs when $\sqrt{\tau_{SF}\tau_D} < T_2^*$. The higher the pump pulse energy is, the shorter τ_D is. Thus, the dephasing is limited by the pump pulse energy. In our experiment, the value of $\sqrt{\tau_{SF}\tau_D}$ = 22 ps (τ_{SF} = 8 ps and τ_D= 62 ps in Figure 2g), which is in the range of T_2^* about 10–50 ps typical for high quality bulk semiconductors [53].

As the excitation fluence further increases, the growth rate of the field becomes high enough to exceed the dephasing rate and SF emission develops. The linewidth initially drops with increasing pump intensity and becomes saturated once SF becomes dominant.

However, in condensed matter systems, SF has been difficult to observe due to the inherently short coherence times of carriers. The SF in $In_{0.2}Ga_{0.8}As/GaAs$ QWs was attributed to many-body renormalization and Coulomb enhancement of gain [54]. Strongly localized states of the subband of $In_{0.1}Ga_{0.9}N/GaN$ QWs have been demonstrated in a few nm carrier localization domains and electrons travel at most over a scale of a few tens of nm [55]. It can be reasonably inferred that the single localized state spontaneously develops into macroscopically coherent emitters due to the high oscillator strength, which devotes to observing SF at room temperature. In the following work, we will study the coherent process from the polarization characteristics of the emission. Under strong excitation, the interaction among multi-bodies in quantum wells deserves research.

4. Conclusions

Our measurements reveal that coherent SF coupling presents in $In_{0.1}Ga_{0.9}N/GaN$ quantum well due to strongly localized states at high excitation fluence and room temperature. This opened up new opportunities to deepen the understanding of the emission mechanism in $In_{0.1}Ga_{0.9}N/GaN$ quantum wells as the prevailing active layer of optoelectronic devices with high-brightness at a higher injected carrier density, and could enable the exploitation of cooperative effects for optoelectronic devices and then prompted us to study the coherent polarization characteristics and the coherent dephasing process of emitters by four-wave-mixing [56], transient resonant four-wave mixing [57], and the second-order coherence of the emission [38]. The interaction among multi-bodies in $In_{0.1}Ga_{0.9}N/GaN$ quantum wells and other semiconductor quantum wells should still be a research branch due to their wide applications for a large variety of optoelectronic devices.

Author Contributions: Conceptualization, methodology, writing—original draft, writing review and editing, investigation, supervision, C.D.; fomal analysis, Z.L.; data curation, experimental technology, X.Z.; resources, B.Z. All authors have read and agreed to the published version of the manuscript.

Funding: This research received the external funding—National Natural Science Foundation of China (Grant No. 62035016).

Institutional Review Board Statement: Not applicable.

Informed Consent Statement: Not applicable.

Data Availability Statement: Data underlying the results presented in this paper are not publicly available at this time but may be obtained from the authors upon reasonable request.

Acknowledgments: The authors would like to thank Jiang Hao, Shen Bing for fruitful discussions and Dong Jian Wen for helps. The authors gratefully acknowledge the support of the State Key Laboratory of Optoelectronic Materials and Technologies.

Conflicts of Interest: The authors declare no conflict of interest.

References

1. Brasser, C.; Bruckbauer, J.; Gong, Y.; Jiu, L.; Bai, J.; Warzecha, M.; Edwards, P.R.; Wang, T.; Martin, R.W. Cathodoluminescence studies of chevron features in semi-polar $(11\bar{2}2)$InGaN/GaN multiple quantum well structures. *J. Appl. Phys.* **2018**, *123*, 174502–174506. [CrossRef]
2. Ryu, H.Y.; Ryu, G.H.; Onwukaeme, C.; Ma, B. Temperature dependence of the Auger recombination coefficient in InGaN/GaN multiple-quantum-well light-emitting diodes. *Opt. Express* **2020**, *28*, 27459–27472. [CrossRef] [PubMed]
3. Chow, Y.C.; Lee, C.; Wong, M.S.; Wu, Y.R.; Nakamura, S.J.; Denbaars, S.P.; Bowers, J.E.; Speck, J.S. Dependence of carrier escape lifetimes on quantum barrier thickness in InGaN/GaN multiple quantum well photodetectors. *Opt. Express* **2020**, *28*, 23796–23805. [CrossRef] [PubMed]
4. Kang, C.H.; Liu, G.; Lee, C.; Alkhazragi, O.; Wagstaff, J.M.; Li, K.H.; Alhawaj, F.; Ng, T.K.; Speck, J.S.; Nakamura, S.; et al. Semipolar $(20\bar{2}\bar{1})$ InGaN/GaN micro-photodetector for gigabit-per-second visible light communication. *Appl. Phys. Express* **2020**, *13*, 014001–014005. [CrossRef]
5. Tanner, D.S.P.; McMahon, J.M.; Schulz, S. Interface Roughness, Carrier Localization, and Wave Function Overlap in c-Plane (In,Ga)N/GaN Quantum Wells: Interplay of Well Width, Alloy Microstructure, Structural Inhomogeneities, and Coulomb Effects. *Phys. Rev. Appl.* **2018**, *10*, 034027–034045. [CrossRef]
6. Christian, G.M.; Schulz, S.; Kappers, M.J.; Humphreys, C.J.; Oliver, R.A.; Dawson, P. Recombination from polar InGaN/GaN quantum well structures at high excitation carrier densities. *Phys. Rev. B* **2018**, *98*, 155301–155309. [CrossRef]
7. Poltavtsev, S.V.; Solovev, I.A.; Akimov, I.A.; Chaldyshev, V.V.; Lundin, W.V.; Sakharov, A.V.; Tsatsulnikov, A.F.; Yakovlev, D.R.; Bayer, M. Long coherent dynamics of localized excitons in (In,Ga)N/GaN quantum wells. *Phys. Rev. B* **2018**, *98*, 195315–195320. [CrossRef]
8. Kazazis, S.A.; Papadomanolaki, E.; Iliopoulos, E. Tuning carrier localization in In-rich InGaN alloys: Correlations between growth kinetics and optical properties. *J. Appl. Phys.* **2020**, *127*, 225701–225711. [CrossRef]
9. Chaudhuri, D.; Kelleher, J.C.; O'Brien, M.R.; O'Reilly, E.P.; Schulz, S. Electronic structure of semiconductor nanostructures: A modified localization landscape theory. *Phys. Rev. B* **2020**, *101*, 035430–035441. [CrossRef]
10. Suski, T.; Staszczak, G.; Korona, K.P.; Lefebvre, P.; Monroy, E.; Drozdz, P.A.; Muzioł, G.; Skierbiszewski, C.; Kulczykowski, M.; Matuszewski, M.; et al. Switching of exciton character in double InGaN/GaN quantum wells. *Phys. Rev. B* **2018**, *98*, 165302–165313. [CrossRef]
11. Vito, A.D.; Pecchia, A.; Carlo, A.D.; Auf der Maur, M. Impact of Compositional Nonuniformity in (In,Ga)N-Based Light-Emitting Diodes. *Phys. Rev. Appl.* **2019**, *12*, 014055–014061. [CrossRef]
12. Aleksiejūnas, R.; Nomeika, K.; Kravcov, O.; Nargelas, S.; Kuritzky, L.; Lynsky, C.; Nakamura, S.; Weisbuch, C.; Speck, J.S. Impact of Alloy-Disorder-Induced Localization on Hole Diffusion in Highly Excited c-Plane and m-Plane (In,Ga)N Quantum Wells. *Rev. Appl.* **2020**, *14*, 054043–054053. [CrossRef]
13. Callsen, G.; Butté, R.; Grandjean, N. Probing Alloy Formation Using Different Excitonic Species: The Particular Case of InGaN. *Phys. Rev.* **2019**, *9*, 031030–031049. [CrossRef]
14. Pashartis, C.; Rubel, O. Localization of Electronic States in III-V Semiconductor Alloys: A Comparative Study. *Phys. Rev. Appl.* **2017**, *7*, 064011–064022. [CrossRef]
15. Anderson, P.W. Absence of Diffusion in Certain Random Lattices. *Phys. Rev.* **1958**, *109*, 1492–1505. [CrossRef]
16. Eliseev, P.G.; Perlin, P.; Lee, J.; Osiński, M. Blue temperature-induced shift and band-tail emission in InGaN-based light sources. *Appl. Phys.Lett.* **1997**, *71*, 569–571. [CrossRef]
17. Cho, Y.-H.; Gainer, G.H.; Fischer, A.J.; Song, J.J.; Keller, S.; Mishra, U.K.; DenBaars, S.P. "S-shaped" temperature-dependent emission shift and carrier dynamics in InGaN/GaN multiple quantum wells. *Appl. Phys. Lett.* **1998**, *73*, 1370–1372. [CrossRef]
18. Dawson, P.; Schulz, S.; Oliver, R.A.; Kappers, M.J.; Humphreys, C.J. The nature of carrier localisation in polar and nonpolar InGaN/GaN quantum wells. *J. Appl. Phys.* **2016**, *119*, 181505. [CrossRef]
19. Feng, S.W.; Liao, P.-H.; Leung, B.; Han, J.; Yang, F.-W.; Wang, H.-C. Efficient carrier relaxation and fast carrier recombination of N-polar InGaN/GaN light emitting diodes. *J. Appl. Phys.* **2015**, *118*, 043104–043108. [CrossRef]
20. Wang, T.; Bai, J.; Sakai, S.; Ho, J.K. Investigation of the emission mechanism in InGaN/GaN-based light-emitting diodes. *Appl. Phys. Lett.* **2001**, *78*, 2617–2619. [CrossRef]
21. Alam, S.; Sundaram, S.; Elouneg-Jamroz, M.; Li, X.; Gmili, Y.E.; Robin, I.C.; Voss, P.L.; Salvestrini, J.-P.; Ougazzaden, A. InGaN/InGaN multiple-quantum-well grown on InGaN/GaN semi-bulk buffer for blue to cyan emission with improved optical emission and efficiency droop. *Superlattice. Microst.* **2017**, *104*, 291–297. [CrossRef]
22. Damilano, B.; Gil, B. Yellow–red emission from (Ga,In)N heterostructures. *J. Phys. D Appl. Phys.* **2015**, *48*, 403001. [CrossRef]

23. Ponce, F.A.; Bour, D.P. Nitride-based semiconductors for blue and green light-emitting devices. *Nature* **1997**, *386*, 351–359. [CrossRef]
24. Wu, F.; Lin, Y.D.; Chakraborty, A.; Ohta, H.; DenBaars, S.P.; Nakamura, S.; Speckv, J.S. Stacking fault formation in the long wavelength InGaN/GaN multiple quantum wells grown on -plane GaN. *Appl. Phys. Lett.* **2010**, *96*, 231912. [CrossRef]
25. Takeuchi, T.; Sota, S.; Katsuragawa, M.; Komoriv, M.; Takeuchi, H.; Amano, H.; Akasaki, I. Quantum-Confined Stark Effect due to Piezoelectric Fields in GaInN Strained Quantum Wells. *Jpn. J. Appl. Phys.* **1997**, *36*, L382–L385. [CrossRef]
26. Wang, T.; Parbrook, P.J.; Fan, W.H.; Fox, A.M. Optical investigation of multiple-quantum wells under high excitation. *Appl. Phys. Lett.* **2004**, *84*, 5159–5161. [CrossRef]
27. Sun, G.; Xu, G.; Ding, Y.J.; Zhao, H.; Liu, G.; Zhang, J.; Tansu, N. Investigation of fast and slow decays in InGaN/GaN quantum wells. *Appl. Phys. Lett.* **2011**, *99*, 081104. [CrossRef]
28. Schulz, T.; Nirschl, A.; Drechsel, P.; Nippert, F.; Markurt, T.; Albrecht, M.; Hoffmann, A. Recombination dynamics in $InxGa_{1-x}N$ quantum wells—Contribution of excited subband recombination to carrier leakage. *Appl. Phys. Lett.* **2014**, *105*, 181109. [CrossRef]
29. Xing, Y.; Zhao, D.; Jiang, D.; Liu, Z.; Zhu, J.; Chen, P.; Yang, J.; Liang, F.; Liu, S.; Zhang, L. Carrier Redistribution Between Two Kinds of Localized States in the InGaN/GaN Quantum Wells Studied by Photoluminescence. *Nanoscale Res. Lett.* **2019**, *14*, 88–95. [CrossRef]
30. Wang, T.; Parbrook, P.J.; Whitehead, M.A.; Fanc, W.H.; Fox, A.M. Study of stimulated emission from InGaN/GaN multiple quantum well structures. *J. Cryst. Growth* **2004**, *273*, 48–53. [CrossRef]
31. Ichimiya, M.; Watanabe, M.; Ohata, T.; Hayashi, T.; Ishibashi, A. Effect of uniaxial stress on photoluminescence in GaN and stimulated emission in $In_xGa_{1-x}N$/GaN multiple quantum wells. *Phys. Rev. B* **2003**, *68*, 035328–035334. [CrossRef]
32. Minj, A.; Romero, M.F.; Wang, Y.; Tuna, Ö.; Feneberg, M.; Goldhahn, R.; Schmerber, G.; Ruterana, P.; Giesen, C.; Heuken, M. Stimulated emission via electron-hole plasma recombination in fully strained single InGaN/GaN heterostructures. *Appl. Phys. Lett.* **2016**, *109*, 221106–221110. [CrossRef]
33. Wang, Z.; Wang, L.; Xing, Y.; Yang, D.; Yu, J.; Hao, Z.; Sun, C.; Xiong, B.; Han, Y.; Wang, J.; et al. Consistency on Two Kinds of Localized Centers Examined from Temperature-Dependent and Time-Resolved Photoluminescence in InGaN/GaN Multiple Quantum Wells. *ACS Photonics* **2017**, *4*, 2078–2084. [CrossRef]
34. Morel, A.; Lefebvre, P.; Kalliakos, S.; Taliercio, T.; Bretagnon, T.; Gil, B. Donor-acceptor-like behavior of electron-hole pair recombinations in low-dimensional (Ga,In)N/GaN systems. *Phys. Rev. B* **2003**, *68*, 045331–045336. [CrossRef]
35. Brosseau, C.-N.; Perrin, M.; Silva, C.; Leonelli, R. Carrier recombination dynamics in $In_xGa_{1-x}N$/GaN multiple quantum wells. *Phys. Rev. B* **2010**, *82*, 085305–085309. [CrossRef]
36. Dicke, R.H. Coherence in Spontaneous Radiation Processes. *Phys. Rev.* **1954**, *93*, 99–110. [CrossRef]
37. Shribanowitz, N.; Herman, I.P.; MacGillivray, J.C.; Feld, M.S. Observation of Dicke Superradiance in Optically Pumped HF Gas. *Phys. Rev. Lett.* **1973**, *30*, 309–312. [CrossRef]
38. Rainò, G.; Becker, M.A.; Bodnarchuk, M.I.; Mahrt, R.F.; Kovalenko, M.V.; Stöferle, T. Superfluorescence from lead halide perovskite quantum dot superlattices. *Nature* **2018**, *563*, 671–675. [CrossRef]
39. Dai, D.C.; Monkman, A.P. Observation of superfluorescence from a quantum ensemble of coherent excitons in a ZnTe crystal: Evidence for spontaneous Bose-Einstein condensation of excitons. *Phys. Rev. B* **2011**, *84*, 115206–115213. [CrossRef]
40. Cong, K.; Zhang, Q.; Wang, Y.; Noe, G.T.; Belyanin, A.; Kono, J. Dicke superradiance in solids [Invited]. *J. Opt. Soc. Am. B* **2016**, *33*, C80. [CrossRef]
41. Bonifacio, R.; Lugiato, L.A. Cooperative radiation processes in two-level systems: Superfluorescence. *Phys. Rev. A* **1975**, *11*, 1507–1521. [CrossRef]
42. Heinzen, D.J.; Thomas, J.E.; Feld, M.S. Coherent Ringing in Superfluorescence. *Phys. Rev. Lett.* **1985**, *54*, 677–680. [CrossRef] [PubMed]
43. Kuraptsev, A.S.; Sokolov, I.M. Many-body cooperative effects in an ensemble of pointlike impurity centers near a charged conductive surface. *Phys. Rev. A* **2019**, *100*, 063836. [CrossRef]
44. Kuokstis, E.; Yang, J.W.; Simin, G.; Khan, M.A.; Gaska, R.; Shur, M.S. Two mechanisms of blueshift of edge emission in InGaN-based epilayers and multiple quantum wells. *Appl. Phys. Lett.* **2002**, *80*, 977–979. [CrossRef]
45. Aleksiejunas, R.; Gelzinyte, K.; Nargelas, S.; Jarasiunas, K.; Vengris, M.; Armour, E.A.; Byrnes, D.P.; Arif, R.A.; Lee, S.M.; Papasouliotis, G.D. Diffusion-driven and excitation-dependent recombination rate in blue InGaN/GaN quantum well structures. *Appl. Phys. Lett.* **2019**, *104*, 022114–022117. [CrossRef]
46. Holmes, M.J.; Choi, K.; Kako, S.; Arita, M.; Arakawa, Y. Room-Temperature Triggered Single Photon Emission from a III-Nitride Site-Controlled Nanowire Quantum Dot. *Nano Lett.* **2014**, *14*, 982–986. [CrossRef]
47. Broeck, D.D.; Bharrat, D.; Hosalli, A.M.; El-Masry, N.A.; Bedair, S.M. Strain-balanced InGaN/GaN multiple quantum wells. *Appl. Phys. Lett.* **2014**, *105*, 031107. [CrossRef]
48. Xu, G.; Sun, G.; Ding, Y.J.; Zhao, H.; Liu, G.; Zhang, J.; Tansu, N. Investigation of large Stark shifts in InGaN/GaN multiple quantum wells. *J. Appl. Phys.* **2013**, *113*, 033104–044108. [CrossRef]
49. Carvalho, L.C.; Schleife, A.; Bechstedt, F. Influence of exchange and correlation on structural and electronic properties of AlN, GaN, and InN polytypes. *Phys. Rev. B* **2011**, *84*, 195105. [CrossRef]
50. Kim, J.H.; Noe, G.T., II; McGill, S.A.; Wang, Y.; Wójcik, A.K.; Belyanin, A.A.; Kono, J. Fermi-edge superfluorescence from a quantum-degenerate electron-hole gas. *Sci. Rep.* **2013**, *3*, 3283–3289. [CrossRef]

51. Mattar, F.P.; Gibbs, H.M.; McCall, S.L.; Feld, M.S. Transverse Effects in Superfluorescence. *Phys. Rev. Lett.* **1981**, *46*, 1123–1126. [CrossRef]
52. Schuurmans, M.F.H.; Vrehen, Q.H.F.; Polder, D. Superfluorescence. *Adv. Atom. Mol. Phys.* **1981**, *17*, 167–228.
53. Klingshirn, C.F. *Semiconductor Optics*; Springer: Berlin/Heidelberg, Germany, 1997; Chapter 22; p. 339.
54. Belyani, A.; Kono, J. Superfluorescence from photoexcited semiconductor quantum wells: Magnetic field, temperature, and excitation power dependence. *Phys. Rev. B* **2015**, *91*, 235448–235457.
55. Hahn, W.; Lentali, J.-M.; Polovodov, P.; Young, N.; Nakamura, S.; Speck, J.S.; Weisbuch, C.; Filoche, M.; Wu, Y.-R.; Piccardo, M.; et al. Evidence of nanoscale Anderson localization induced by intrinsic compositional disorder in InGaN/GaN quantum wells by scanning tunneling luminescence spectroscopy. *Phys. Rev. B* **2018**, *98*, 045305–045309. [CrossRef]
56. Steel, D.G.; Wang, H. Dephasing of optically induced excitonic coherences in semiconductor Heterostructures. *Appl. Phys. A* **2000**, *71*, 519–524. [CrossRef]
57. Poltavtsev, S.V.; Kosarev, A.N.; Akimov, I.A.; Yakovlev, D.R.; Sadofev, S.; Puls, J.; Hoffmann, S.P.; Albert, M.; Meier, C.; Meier, T.; et al. Time-resolved photon echoes from donor-bound excitons in ZnO epitaxial layers. *Phys. Rev. B* **2017**, *96*, 035203–035207. [CrossRef]

nanomaterials

MDPI

Review

Recent Progress in Contact Engineering of Field-Effect Transistor Based on Two-Dimensional Materials

Jialei Miao [1,2], Xiaowei Zhang [1,*], Ye Tian [1] and Yuda Zhao [2,3,*]

[1] Department of Electrical Engineering and Computer Science, Ningbo University, Ningbo 315211, China
[2] School of Micro-Nano Electronics, Hangzhou Global Scientific and Technological Innovation Centre, Zhejiang University, 38 Zheda Road, Hangzhou 310027, China
[3] Key Laboratory of Optoelectronic Chemical Materials and Devices of Ministry of Education, Jianghan University, Wuhan 430056, China
* Correspondence: zhangxiaowei@nbu.edu.cn (X.Z.); yudazhao@zju.edu.cn (Y.Z.)

Abstract: Two-dimensional (2D) semiconductors have been considered as promising candidates to fabricate ultimately scaled field-effect transistors (FETs), due to the atomically thin thickness and high carrier mobility. However, the performance of FETs based on 2D semiconductors has been limited by extrinsic factors, including high contact resistance, strong interfacial scattering, and unintentional doping. Among these challenges, contact resistance is a dominant issue, and important progress has been made in recent years. In this review, the Schottky–Mott model is introduced to show the ideal Schottky barrier, and we further discuss the contribution of the Fermi-level pinning effect to the high contact resistance in 2D semiconductor devices. In 2D FETs, Fermi-level pinning is attributed to the high-energy metal deposition process, which would damage the lattice of atomically thin 2D semiconductors and induce the pinning of the metal Fermi level. Then, two contact structures and the strategies to fabricate low-contact-resistance short-channel 2D FETs are introduced. Finally, our review provides practical guidelines for the realization of high-performance 2D-semiconductors-based FETs with low contact resistance and discusses the outlook of this field.

Keywords: contact resistance; two-dimensional (2D) materials; Fermi-level pinning; transistor

1. Introduction

The development of silicon complementary metal–oxide semiconductor (CMOS) integrated circuits has followed Moore's law for several decades. With the downscaling of the transistor dimensions into sub−20 nanometer nodes, two main challenges emerge, including drain-induced barrier lowering and degradation of the carrier mobility of Si. The former effect makes the transistor hard to turn off and results in a high off-current, which will cause high static power consumption. Emerging technologies such as FinFETs and Gate-All-Around (GAA) FETs have been employed to decrease the off-current. The mobility degradation is caused by the strong interfacial scattering, especially when the semiconductor thickness (t_b) is in the sub−3−nm regime [1]. Since the discovery of graphene, two-dimensional (2D) materials with atomic thickness exhibit high carrier mobility, even at $t_b < 3$ nm [1], and show high immunity to the short-channel effect. These unique properties contribute to making 2D materials promising candidates to fabricate ultimately scaled transistors.

Although 2D materials present a unique pathway to build next-generation electronic devices, the construction of 2D-materials-based transistors (2D FETs) faces several technical challenges, including the growth of high-quality wafer-scale 2D materials [2], wafer-scale transfer methods [3], low contact resistance [4], and the high-quality dielectric interface [5]. These technical issues lead to large device-to-device variation [6] and the apparent discrepancy between the theoretical prediction and actual device performance, limiting the industrial applications of 2D materials in logic devices. Among these challenges, the contact problems are of vital important because the working mechanism of 2D FETs is based on the control of charge injection at the metal/2D junction, which is quite different from silicon CMOS transistors [4]. Recent studies show tremendous advances in the achievement of the ideal Mott–Schottky contact and the lowering of the contact resistance in 2D FETs [7–9]. The value of the contact resistance in 2D-materials-based transistors is approaching the requirement of the International Roadmap for Devices and Systems (IRDS) 2024 targets of logic transistors [10]. It is believed that a summary of emerging strategies to realize contact engineering in 2D FETs is urgently needed.

In this review, we present a comprehensive analysis of contact challenges in 2D FETs and discuss the recent research progress. We start with the origins of high contact resistance in 2D FETs. Then, two contact structures are presented, including top contact and edge contact, followed by several strategies to decrease the contact resistance. Finally, an outlook is provided to present the possible roadmap for the contact engineering of 2D FETs.

2. Fermi-Level Pinning

2.1. Fermi-Level Pinning and Pinning Factor

The Schottky barrier height (Φ_{SB}) and the contact resistance (R_c) are important quantitative parameters to examine the quality of the metal–2D material junction. In an ideal metal–semiconductor junction, Φ_{SB} is determined by the Schottky–Mott rule based on the energy-level band alignment [11,12]:

$$\Phi_{SB,n} = \Phi_M - \chi_s \tag{1}$$

$$\Phi_{SB,p} = I_s - \Phi_M \tag{2}$$

where $\Phi_{SB,n}$ and $\Phi_{SB,p}$ are the Schottky barrier heights for electrons and holes transport, Φ_M represents the metal work function, χ_s represents the electron affinity and I_s represents the ionization potential of the semiconductor. The Φ_{SB} is linearly dependent on the metal work function in the Schottky–Mott model, as shown in Figure 1a. However, the metal work function in FETs is always derived from the theoretical value, and it is pinned on a specific position within the bandgap of the semiconductor regardless of the selection of different metals, as shown in Figure 1b. This effect is called Fermi-level pinning, which makes the metal–semiconductor junction insensitive to the modulation of the metal work function. The pinning factor S represents the strength of Fermi-level pinning:

$$S = |d\Phi_{SB}/d\Phi_M| \tag{3}$$

The value of S in an ideal device is nearly equal to 1, but S is usually far away from 1 in 2D FETs. Liu et al. and Kim et al. demonstrated the S value of ~0.1 in MoS_2 FETs with deposited metal contact [13,14]. The Fermi-level pinning effect strongly limits the performance of 2D semiconductor FETs.

2.2. Origins of Fermi-Level Pinning

Tersoff et al. successfully established a parameter-free metal-induced gap states (MIGS) model in bulk semiconductors to explain the Fermi-level pinning effect. The MIGS model can quantitatively explain the almost unchanged Schottky barrier height, which is independent of the metal work function in experiments [15]. Guo et al. employed the density functional theory (DFT) calculation to calculate the Schottky barrier height of 2D transition metal dichalcogenides (TMDs) by using different metal contacts. The calculated pinning factor is around 0.3, demonstrating a strong Fermi-level pinning effect. They found that direct bonding existed between the contact metal atoms and the chalcogen atoms of TMDs, leading to the MIGS [16]. The charge neutrality level serves as a quantitative characteristic of the electronic states of the defective semiconductor surface. Dominik et al. employed the primary theoretical model to calculate the charge neutrality level (CNL) of the monolayer TMDs, and they found that CNL is mostly placed near the mid-point of the semiconducting band gaps [17]. Although they use different theoretical calculation methods, both studies demonstrate that the MIGS can well-explain the Fermi-level pinning effect in 2D TMDs. Fermi-level pinning in 2D FETs mainly originates from the interfacial states. In the early stage of the study, Au, Ni and Pt with a high melting temperature are selected as the metal contact of 2D FETs, and they are deposited by evaporation or sputtering techniques, as shown in Figure 1a,b. This leads to the compact stacking of metal atoms on the surface of 2D layers, the wavefunction interaction between the metal and 2D semiconductor, and the rehybridizations of the semiconductor's original wavefunctions, resulting in the strong orbital overlap and MIGS. Photolithography and electron-beam lithography are common techniques to pattern the electrodes on 2D materials. Compared with bulk semiconductors, atomically thin 2D materials are sensitive to laser [18], electron-beam [19,20] and chemical solution [21]. Matsunage et al. reported that a relatively low electron-beam dose (280 $\mu C/cm^2$) used in conventional electron-beam lithography will induce strain in MoS_2, leading to the local widening of the MoS_2 bandgap [22]. Preeti et al. systemically reported the doping effect of the conventional lithography process and the used chemical solvent. For example, acetone shows n-type doping and chloroform displays p-type doping on MoS_2 [23]. Meanwhile, the high-energy deposition process generates atomic defects at the interface of metal and 2D layers, facilitates the formation of covalent bonds [24,25], and gives rise to MIGS. Liu et al. experimentally showed that the typical metal deposition process induced defects at the contact region, which were observed by transmission electron microscope [13]. In 2D FETs with deposited metal contact, the injected charges are accumulated at the interfacial gap states regardless of the modulation of the metal work function, as shown in Figure 1c. The Fermi level is pinned around these gap states and a Schottky barrier is unavoidable, as shown in Figure 1d. Furthermore, heavy doping via ion implantation is employed in Si CMOS FETs to realize Fermi-level depinning and successfully achieve low contact resistance. However, the ion implantation cannot be well implemented on 2D materials. The implantation process will generate a large number of defects in atomically thin 2D materials and degrade the carrier transport.

Overall, the interface defects of the 2D semiconductor can induce the MIGS and greatly affect the contact quality. Furthermore, the conventional Fermi-level depinning methods in Si CMOS FETs are not suitable for 2D FETs. Therefore, the specific contact strategies should be tailored to meet the requirement of 2D FETs.

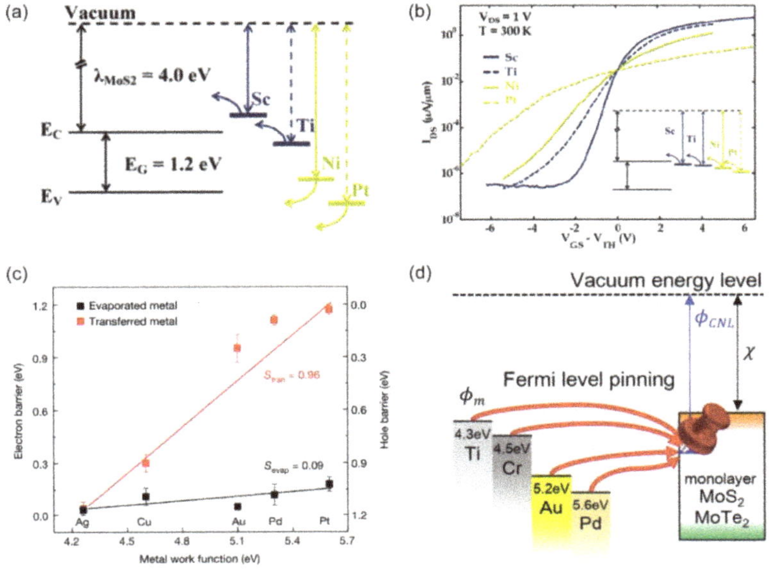

Figure 1. Origin and influence of Fermi-level pinning. (**a**) Expected metal Fermi level with the electronic bands of MoS_2 and different metal electrodes with different work function. (**b**) Transfer curves with different metals. The inset shows the actual line-up based on the experimental data. Reprinted/adapted with permission from Ref. [26]. Copyright 2012, American Chemical Society. (**c**) Experimentally determined Schottky barrier height for transfer of different transfer metals and evaporated metals. Reprinted/adapted with permission from Ref. [13]. Copyright 2018, Springer Nature. (**d**) Schematic illustration of Fermi-level pinning. Reprinted/adapted with permission from Ref. [27]. Copyright 2017, American Chemical Society.

3. Contact Engineering of 2D FETs

FETs based on 2D materials can realize the ultimate downscaling of transistor dimensions. Along with the scaling of channel length, the scaling of contact region will produce new challenges. The most straightforward effect is the increase in contact resistance in the scaled contact region. The transfer length is the effective length with which carriers are transferred from the contact metal to channels. According to the current crowding model [28], carriers prefer to be transferred inside the more conductive metal and enter the semiconductor only near the end of metal–semiconductor contact regions, leading to a much smaller transfer length in comparison with the contact length. In this condition, the contact resistance is dependent on the transfer length, rather than the contact length, and its value can be calculated by the distributed resistor network model [29]. When the contact length is equal to or smaller than the transfer length, the charge injection is limited by the contact length. Therefore, a high-quality metal–semiconductor contact becomes more important in 2D FETs for ultra-scaled integrated circuits.

In order to overcome the strong Fermi-level pinning effect in 2D FETs, great progress has been made in the optimized metal-deposition process and the efficient carrier-modulation methods. In this section, state-of-the-art contact engineering strategies are systematically discussed.

3.1. Top Contact Engineering

Top contact refers to the deposition of metal contact on the surface of 2D layers. Due to the large surface-to-edge ratio, top contact is widely used in 2D FETs, and the fabrication process is simple and efficient. In this part, we introduce the use of bulk metals, bulk semimetals and 2D metals as top contact to optimize the contact of 2D FETs.

3.1.1. Bulk Metals

Defects in 2D materials, including intrinsic defects and the generated defect during the fabrication process, are the main origins of the Fermi-level pinning effect. Defects contribute to the interfacial trap states, which is directly related to the Fermi-level pinning effect. In early studies, bulk metals with low work function were used to achieve high-performance n-type 2D FETs. Kwon et al. reported the use of Al as a contact metal in MoS_2 FETs and obtained a Schottky barrier height of 70 meV [30]. Das et al. reported the use of Sc as a contact metal in MoS_2 FETs [26] due to the low work function of Sc (3.5 eV). An average Schottky barrier of 0.03 eV was achieved, which represents a very low contact barrier at that time. However, from the Schottky–Mott rule, the ideal Schottky barrier between Sc and MoS_2 should be negative because the work function of Sc is much lower than the CBM of MoS_2. The positive Schottky barrier experimentally demonstrates Fermi-level pinning. In order to avoid the generation of defects during the metal-deposition process, transferred metal methods have been developed to preserve a sharp and clean interface between metal and 2D layers. Liu et al. reported realization of the van der Waals (vdW) contact through mechanically transferring metals to avoid chemical bonding and defect-induced gap states, as shown in Figure 2a,b. Owing to the atomically clean interface between metal and semiconductor, the majority carrier type and corresponding Schottky barrier height are strongly dependent on the metal work function ($S = 0.96$) [13]. Wang et al. further reported that the transferred metal can achieve low contact resistance in p-type WSe_2 FETs [31]. Due to the successful Fermi-level depinning, the work function of the transferred metal plays an important role in determining the charge injection and the device performance. Han et al. reported the use of molecular functionalization to change the work function of gold electrodes. Then, they fabricated top-contact FETs via the transfer of these pre-modified electrodes to tune the charge injection in MoS_2 FETs [32], demonstrating the modulation of the Schottky barrier. This method has also been used in 2D-materials-based resistive random-access memory, leading to a stable resistive switching performance [33]. Therefore, the transferred metal method represents a reliable way to create vdW contact [34], but wafer-scale metal transfer technology is still lacking. Moreover, the metal transfer process is mostly operated under the optical microscope and the alignment error is still huge, which is another big challenge.

The conventional metal-deposition process can be optimized to realize low contact resistance. Chris et al. reported that Au deposited in ultra-high vacuum (~10^{-9} Torr) yields three times lower contact resistance than that in normal conditions, as shown in Figure 2c [35]. Wang et al. reported high performance p-type FETs based on single− and few−layer MoS_2 and WSe_2 by the electron-beam evaporation of high-work-function metals such as Pd and Pt, as shown in Figure 2d [8]. They conducted the metal evaporation at a high vacuum ($<10^{-8}$ torr) and low temperature (18 °C to 36 °C) to avoid high temperature damage to 2D semiconductors and to form vdW contact between the metal and semiconductor interface. Wang et al. reported the employment of In metal to build van der Waals contact with 2D semiconductors [36]. Due to the low melting point of In, the temperature of the 2D sample can be greatly decreased and a high-quality vdW interface can form, resulting in the Fermi-level depinning. Furthermore, In metal can be used to form stable alloys with other metals to modulate the work function. The use of In alloy as contact shows the advantages in the preservation of a high-quality contact interface and the effective tunability of the Schottky barrier. Kumar et al. employed In/Au alloy and Sn/Au alloy as contact electrodes in monolayer MoS_2 FETs, and they achieved an ultra-low contact resistance of 190 Ω·μm for In/Au alloy and 270 Ω·μm for Sn/Au alloy [37]. The use of metal alloys increases the thermal stability of low-melting-point metals and results in ~450 °C temperature tolerance that is compatible with back-end-of-line (BEOL).

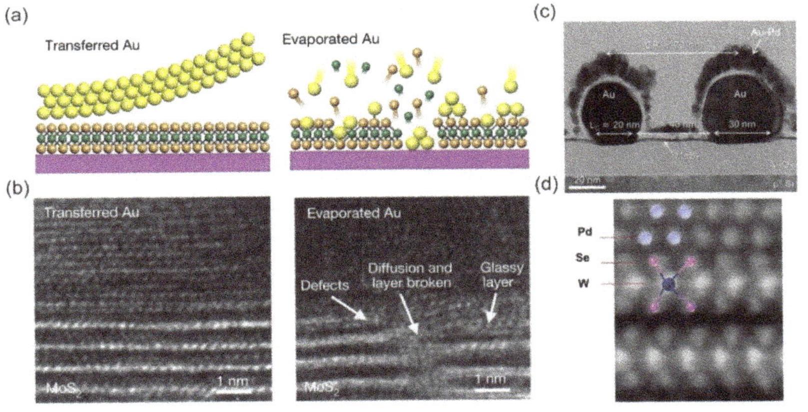

Figure 2. Atomical clean interface is significant to Fermi-level depinning. (**a**) cross-sectional schematics of transferred Au electrodes with atomically sharp and clean metal–semiconductor interfaces and conventional electron beam-deposited Au electrodes with poor interface. (**b**) Cross-section TEM images of transferred Au electrodes and evaporated Au electrodes. Reprinted/adapted with permission from Ref. [13]. Copyright 2018, Springer Nature. (**c**) TEM cross-section of a MoS_2 FET with gold electrodes deposited under ultra-high vacuum. Reproduced with permission from Ref. [35]. Copyright 2016, American Chemical Society. (**d**) cross-sectional STEM of the Pd–WSe_2 interface. Reprinted/adapted with permission from Ref. [8]. Copyright 2022, Springer Nature.

3.1.2. Bulk Semimetals

Recently, a powerful strategy was demonstrated by adopting semi-metals as contact electrodes to suppress MIGS and the Fermi-level pinning effect [7,38]. Shen et al. used semi-metal Bi as the n-type contact metal in monolayer MoS_2 FETs [7]. Bi as a semi-metal has a negligible density of state at the Fermi level, and this induces the suppression of MIGS, as shown in Figure 3a,b. Furthermore, the use of Bi contact results in the degenerately doped MoS_2 with a high electron density of 1.5×10^{13} cm^{-2}, and the Fermi level shifts from inside the bandgap to above the conduction band minimum. They achieved an ultralow contact resistance of 123 $\Omega \cdot \mu m$ and a high current density of 1135 $\mu A\ \mu m^{-1}$ in a 35−nm channel length MoS_2 FETs, as shown in Figure 3c,d. Owing to the highly efficient carrier injection between Bi and MoS_2, the drain current density increased by lowering the temperature from room temperature to 77 K, as shown in Figure 3e. The values of Bi−MoS_2 contact resistance are comparable to those Si transistors and approach the quantum limit, as shown in Figure 3f. However, it has been observed that the MoS_2 FETs with Bi electrodes degraded severely after annealing at 300 and 400 °C [39]. Chou et al. reported semimetal antimony (Sb) as a novel contact metal to enable 2D materials towards advanced electronic device applications. They obtained a near-zero Schottky barrier height and a low contact resistance of 0.66 $k\Omega \cdot \mu m$ [39]. Compared with Bi contact electrodes, the melting point of Sb (630 °C) is much larger than that of Bi (271 °C), although Sb has a higher work function than Bi. The transfer curves of MoS_2 FETs with Sb electrodes show a better electrical performance after high-temperature annealing. Overall, the use of semi-metal as a contact can greatly reduce MIGS and realize Fermi-level depinning in 2D FETs.

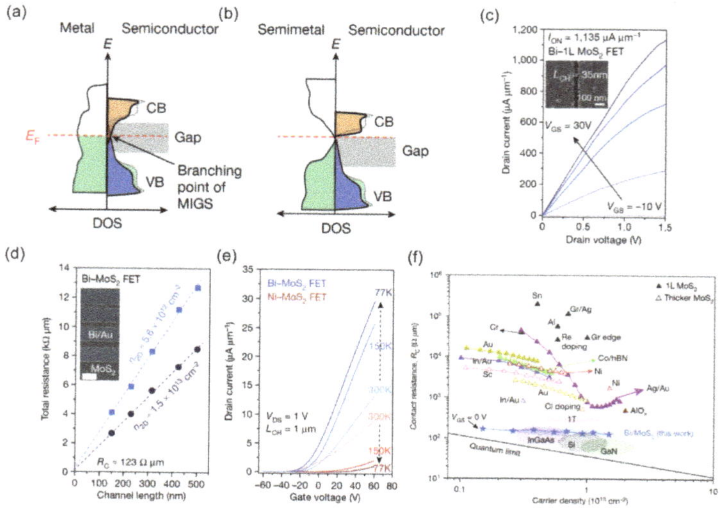

Figure 3. Band structure of normal metal contact (**a**) and bismuth–semiconductor contact (**b**). The Fermi level of the semi-metal aligns with the conduction band of the semiconductor, the density of states at the Fermi level of the semimetal is near-zero, and metal-induced gap states are suppressed. (**c**) Output curves of a 35-nm Bi−MoS_2 FET with a high current density. (**d**) Contact resistance of MoS_2 FETs with bismuth electrodes. (**e**) Transfer curves of Bi−MoS_2 and Ni−MoS_2 FETs at various temperatures. (**f**) State-of-the-art contact technology for MoS_2 transistors as a function of n_{2d}. Reprinted/adapted with permission from Ref. [7]. Copyright 2021, Springer Nature.

3.1.3. Two-Dimensional Metals/Semimetals

The MIGS are commonly found at the interface between 3D metal and 2D semiconductors. Liu et al. theoretically found that the interface states in the metal–semiconductor junction mainly derive from the 3D metal rather than the 2D semiconductor [40]. Therefore, they suggested replacing the 3D bulk metal with 2D metals. In their works, they predict that the Fermi-level pinning effect can be greatly suppressed when the 2D metal–2D semiconductor interface is well formed. The existence of a van der Waals distance between the 2D metal and the 2D semiconductor (3 to 4 Å) leads to the weak interlayer interaction, mild orbital overlap and the creation of interface dipole, contributing to Fermi-level depinning. Two-dimensional layered materials with metallic properties, such as graphene, 1T−MoS_2 and $PtSe_2$, can be used to form vdW contacts on 2D semiconductors. Majumdar et al. employed 2H−$TaSe_2$, graphene and degenerately-doped semiconducting $SnSe_2$ as contact metals [41]. They demonstrated that vdW contacts exhibited a universal Fermi-level depinning phenomenon, as shown in Figure 4a.

We present the theoretical band alignment of MoS_2 and WSe_2 with different 2D metals/semimetals, as shown in Figure 4b [42]. The stacking of 2D metals on 2D semiconductors leads to the ideal Schottky junction or Ohmic contact by selecting 2D metallic materials with suitable work function. Liu et al. also employed graphene as contact electrodes and further encapsulated the MoS_2 channel with top and bottom hexagonal boron nitride (h−BN). The MoS_2 FETs reached a high field-effect mobility of up to 1300 $cm^2V^{-1}s^{-1}$ at a low temperature [43]. Chuang et al. reported the use of $Nb_{0.005}W_{0.995}Se_2$ as contact electrodes and the achievement of a low contact resistance (~0.3 kΩ·μm) [44]. Hwang et al. used chlorine-doped $SnSe_2$ as the high-work-function contact metal in WSe_2 FETs and realized a pronounced p-type Wse$_2$ transistors with the mobility of 15.7 $cm^2V^{-1}s^{-1}$, as shown in Figure 4c [45]. Wu et al. reported the fabrication of bi-layer WSe_2 transistors via the vdW epitaxy and the controlled crack formation processes, as shown in Figure 4d [9]. In a 20 nm−long and 1.3 nm−thick bi-layer WSe_2 transistor, an on-state current density of 1.72 mA μm^{-1} and a contact resistance of 0.25–0.54 k$\Omega \cdot \mu m$ are achieved. $PtSe_2$ has been demonstrated to have a higher electron mobility than MoS_2 based on DFT calculations and experimentally extracted field-effect mobility [46]. Furthermore, $PtSe_2$ shows a layer-dependent semiconductor to semimetal transition. When a $PtSe_2$ transistor is built, few-layer $PtSe_2$ can serve as a semiconducting channel and bulk $PtSe_2$ can serve as the semi-metallic contact [47–49]. Das et al. vertically integrated a thick $PtSe_2$ layer as source/drain contact on the surface of an ultrathin $PtSe_2$ channel, achieving a high performance of all $PtSe_2$ FETs, as shown in Figure 4e [50]. Zhang et al. reported barrier-free p-type WSe_2 FETs with a layered $1T'-WS_2$ semimetal contact, as shown [51]. Owing to the high-quality interface between WSe_2 and $1T'-WS_2$, the WSe_2 FETs achieve a 50 meV Schottky barrier height and a high field-effect mobility of 97 $cm^2V^{-1}s^{-1}$.

The growth of a graphene/MoS_2 heterostructure and the use of graphene as contact have shown the potential to lower the contact resistance of MoS_2 FETs [39]. In Mootheri et al.'s work, they further explored the function of 3D metal in the metal/graphene/MoS_2 contact structure. They proved that Ru–graphene contact show the lowest contact resistance of 9.34 k$\Omega \cdot \mu m$ compared with Pd–graphene and Ni–graphene contact [52].

The use of 2D metallic materials is a simple and effective way to achieve high-quality vdW contact on a 2D semiconductor. However, the stacking of 2D vdW heterostructures needs a complex transfer process during the device fabrication, which is inefficient for the fabrication of large-scale devices. Reliable transfer methods that are suitable for wafer-scale fabrication with a high alignment accuracy need to be explored. Furthermore, it is quite challenging to use the mechanical transfer method to fabricate short-channel devices. The etching of 2D layers with sub−1−micron precision is essential to realize the contact engineering of 2D FETs.

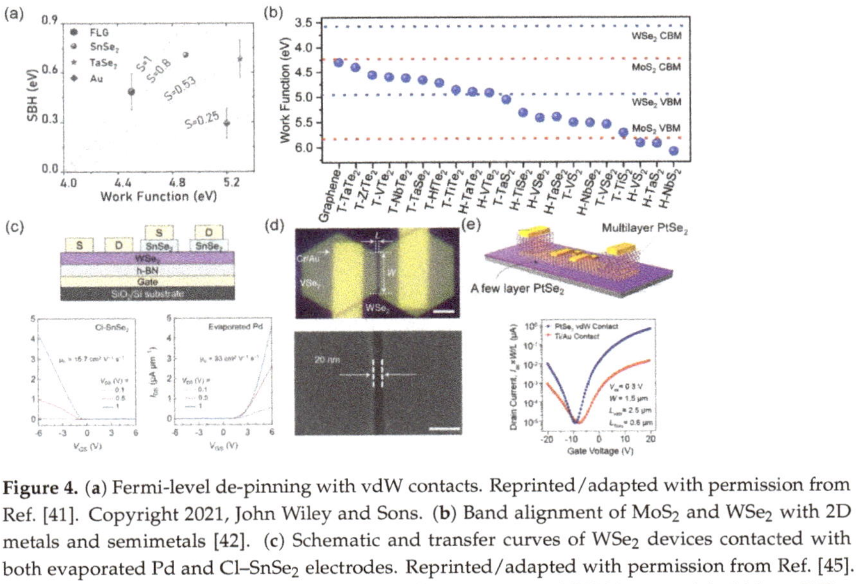

Figure 4. (**a**) Fermi-level de-pinning with vdW contacts. Reprinted/adapted with permission from Ref. [41]. Copyright 2021, John Wiley and Sons. (**b**) Band alignment of MoS_2 and WSe_2 with 2D metals and semimetals [42]. (**c**) Schematic and transfer curves of WSe_2 devices contacted with both evaporated Pd and Cl–$SnSe_2$ electrodes. Reprinted/adapted with permission from Ref. [45]. Copyright 2022, John Wiley and Sons. (**d**) Optical microscopy and SEM images of the bi-layer WSe_2 transistor. Reprinted/adapted with permission from Ref. [9]. Copyright 2022, Springer Nature. (**e**) Schematic illustration of few-layer $PtSe_2$ FETs with metallic $PtSe_2$ vdW contact and a conventional Ti/Au contact. Reprinted/adapted with permission from Ref. [50]. Copyright 2021, American Chemical Society.

3.2. Edge Contact Engineering

Wang et al. first showed the structure of edge contact in 2D FETs by encapsulating a 2D channel with hexagonal boron nitride (h−BN) and exposing the edge of the channel to the metal contact. The edge contact shows several advantages, including being free of Fermi-level pinning induced by interfacial states and having a lower tunnel barrier, strong orbital overlaps, the absence of a Schottky barrier, and high carrier injection efficiency [53]. As the thickness of the 2D layer is very small, effective orbital overlap or hybridization is required between the metal and the edge of the 2D layers, which is the prerequisite to building high-quality edge contact. In monolayer TMDs such as MoS_2 and WSe_2, the CBM arises mainly from the d−orbitals of transition-metal atoms [54]. When a carrier is injected from metal to the conduction bands of MoS_2, the edge contact exhibits strong orbital hybridization with transition-metal atoms [55] and leads to efficient carrier injection. In comparison, the top contact is formed on the surface of chalcogen atoms in monolayer MoS_2 with little influence on the CBM. The realization of edge contact on 2D materials is mainly through plasma etching, metallization and phase engineering. We discuss these methods in this section.

3.2.1. Plasma Etching and Metallization

Jain et al. reported edge-contact monolayer MoS_2 FETs encapsulated with h−BN layers, as shown in Figure 5a [56]. The reactive ion etching, in situ Ar sputtering and annealing removed adsorbate on the MoS_2 surface and preserved the high carrier mobility of ~30 $cm^2V^{-1}s^{-1}$, resulting in a steep subthreshold swing of 116 mV/dec with a negligible hysteresis. Yang et al. reported the polarity control of MoS_2 FETs by employing a 1D elemental metal contact (Figure 5b) [55]. Figure 5c shows the high-resolution transmission electron microscopy (HR−TEM) cross-sectional image of the edge contact area. The use of high-work-function palladium (Pd) or gold (Au) enables a high-quality p-type dominant contact to MoS_2 layers without extrinsic doping, as shown in Figure 5d,e. Moreover, the h−BN encapsulation can suppress the interfacial scattering in 2D FETs and improve the long-term ambient stability, demonstrating the advantages of edge contact structure. Some 2D materials undergo gradual oxidation in air, especially $MoTe_2$, black phosphorus and InSe, which can adopt the edge contact structure [57–59]. Except h−BN encapsulation, other insulating materials such as Al_2O_3 [60] and PMMA [61] have also been used to form edge contacts on 2D semiconductors.

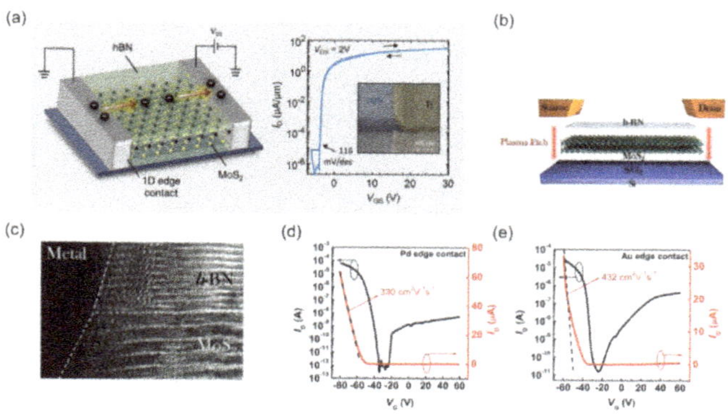

Figure 5. (**a**) Schematic of 1D edge contact MoS_2 FET and transfer curve. Reprinted/adapted with permission from Ref. [56]. Copyright 2019, American Chemical Society. (**b**) Schematic diagram shows 1D edge contact FET with different metals. (**c**) HR−TEM image of Pd−MoS_2 1D edge contact FET; transfer curves of Pd (**d**) and Au (**e**) edge contact MoS_2 FETs, realizing p-type intrinsic MoS_2 FETs and Fermi-level depinning. Reprinted/adapted with permission from Ref. [55]. Copyright 2019, John Wiley and Sons.

3.2.2. Phase Engineering and Degenerate Doping of 2D Layers

Two-dimensional TMDs have been reported with different polymorphs, including hexagonal (2H) and monoclinic or octahedral (1T, 1T′) structures [62–64]. The 2H−phases MoS_2 and WSe_2 show semiconducting properties, while the 1T (1T′) phase displays metallic transport behavior. Therefore, phase engineering between 2H and 1T (1T′) can dramatically change the electronic properties of group−6 TMDs. The transition of group−6 TMDs from 2H to 1T (1T′) phase at the contact region can be used to achieve high-quality edge contact in 2D FETs, which is similar to the degenerate doping at the source/drain region [65–67]. Kappera et al. first demonstrated the phase transition of MoS_2 from 2H to 1T through n-butyllithium treatment, as shown in Figure 6a [68]. The 1T/2H interface dominates the carrier injection, and the device exhibits an ultra-low contact resistance of 200–300 Ω·μm at zero gate bias. However, this 1T−phase MoS_2 is metastable, and the stability is a challenge. This method can be used not only in MoS_2 FETs [69], but also in other 2D-materials-based FETs. Cho et al. reported the laser-induced phase transition of $MoTe_2$ from 2H to 1T′ phase, as shown in Figure 6b [70]. The 1T′ phase region works as the edge contact of the 2H phase channel to improve the carrier injection, and the Schottky barrier height is decreased to 10 meV. They further reported the reversible phase transition of $MoTe_2$ between 2H and 1T′ by controlling the annealing temperature and the cooling speed [71]. The 1T′ $MoTe_2$ has a thermal stability of 300 °C, which is higher than 1T phase MoS_2 [72]. Reversible phase transition of WSe_2 layers has been reported by Ma et al. The n-butyllithium treatment on 2H-phase WSe_2 induces the semiconducting to metallic phase transition, and the thermal annealing drives the metallic phase Wse_2 to be converted back to the semiconducting phase, as shown in Figure 6c [73].

The generation of defects by weak plasma treatment can also Induce phase transition. Zhu et al. reported a facile, clean, controllable and scalable phase-engineering technique for monolayer MoS_2, as shown in Figure 6d [74]. Point defects (single S vacancies) result in the 2H to 1T phase transitions. Akinola et al. also reported a phase transformation in a region of a layered semiconductor $PdSe_2$, as shown in Figure 6e [75]. This phase transition is driven by defects created by argon plasma, and this turns $PdSe_2$ into $Pd_{17}Se_{15}$. Recently, Cai et al. performed plasma treatment on patterned MoS_2 layer to induce a local bonding distortion. This distorted area works as a semi-metallic bridge between the metal and the pristine channel to facilitate the charge injection [76]. The TEM image shows that the distorted MoS_2 displays an octahedral structure, and the device exhibits an ultra-low contact resistance of 90 Ω·μm, approaching the quantum limit.

Another strategy is the introduction of degenerate doping during the growth process. Li et al. reported that unidirectionally aligned monolayer Fe−doped MoS_2 domains are prepared on two-inch commercial c-plane sapphire, suggesting the feasibility of synthesizing wafer-scale-doped 2D semiconductors with outstanding device performance, as shown in Figure 6f [77]. Vu et al. reported a one-step growth approach to synthesize Nb-doped WSe_2 with a controllable doping concentration. The fabricated $NbSe_2$/doped-semiconductor vdW heterostructures have a low contact resistance of 2.46 kΩ·μm [78]. Hemanjaneyulu et al. reported the dramatic n-doping of MoS_2 by immersing it in KI solution. The contact resistance can be greatly reduced to 0.75 kΩ μm [79]. Metallic nanoparticles have also been used to dope a 2D semiconductor channel and further effectively modulate the carrier transport in 2D FETs. Khan et al. reported the charge doping of $ReSe_2$ through the adsorption of Co nanoparticles [80]. Sarkar et al. reported the doping effect of noble metal nanoparticles (Au, Ag, Pd, Pt) in TMDs and revealed the relationship between metal work function and the doping effect in MoS_2 [81].

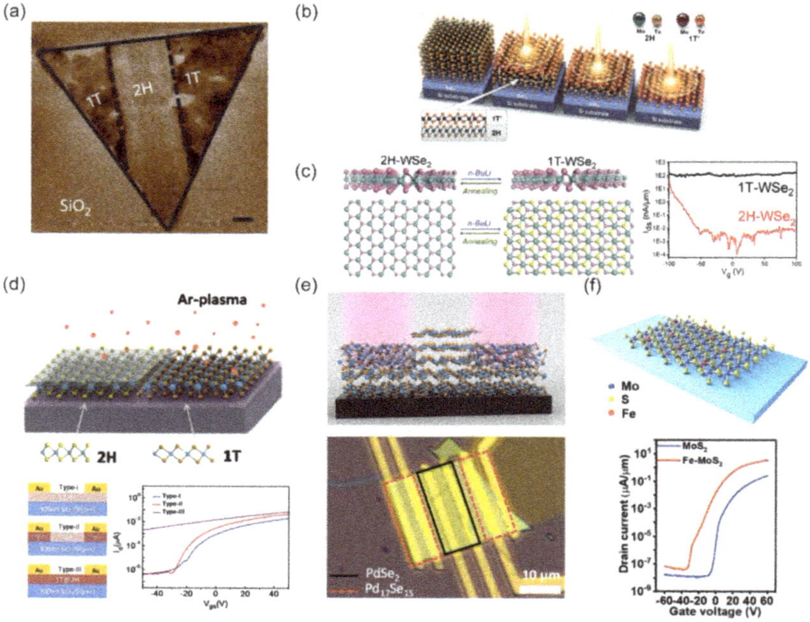

Figure 6. (**a**) Electrostatic force microscopy phase image of a monolayered MoS_2 nanosheet showing the difference between locally patterned 2H and 1T phase. Reprinted/adapted with permission from Ref. [68]. Copyright 2014, Springer Nature. (**b**) Schematic showing laser-irradiation-induced phase transition from 2H to 1T′ Phase in $MoTe_2$. Reprinted/adapted with permission from Ref. [70]. Copyright 2015, American Association for the Advancement of Science. (**c**) Side view and transfer curve comparison of the 2H and 1T phase WSe_2. Reprinted/adapted with permission from Ref. [73]. Copyright 2015, American Chemical Society. (**d**) Schematic formation of 1T phase MoS_2 via plasma treatment and three types of devices. Reprinted/adapted with permission from Ref. [74]. Copyright 2017, American Chemical Society. (**e**) Schematic and optical image of a $PdSe_2$ device with irradiated contact area used to obtain the pristine channel and $Pd_{17}Se_{15}$ contacts. Reprinted/adapted with permission from Ref. [75]. Copyright 2019, American Chemical Society. (**f**) Controllable synthesis and doping determination of monolayer Fe−MoS_2. Reprinted/adapted with permission from Ref. [77]. Copyright 2022, John Wiley and Sons.

3.3. Inserting Interlayer between Metal and 2D Materials

In previous sections, contact engineering has been realized by using 3D or 2D metallic materials, performing phase transition and introducing degenerate doping. Another strategy to suppress the Fermi-level pinning effect is introduced in this section. To decouple the orbital overlap, a thin insulating tunnel layer is inserted between the metal and semiconductor. The insulating buffer layer will increase the distance between the metal and semiconductor, which is an efficient way to suppress interface interaction, and the interlayer will inhibit high energy damage induced by the metal deposition process. The decrease in MIGS results in a reduced Schottky barrier height [82]. However, the thickness of the interlayer should be properly tuned because the electron tunneling through the insulating buffer layer is mandatory. Furthermore, the charges are injected by direct tunneling or Fowler–Nordheim tunneling dependent on the band alignment. Chen et al. first reported the insertion of a thin MgO film for Co−contacted monolayer MoS_2 FETs [83]. The Schottky barrier height was reduced from 60 to 9.7 meV with the increasing MgO thickness from 0.9 to 2 nm. Lee et al. showed a statistical study of Schottky barrier height by inserting a thin tunneling Ta_2O_5 layer between MoS_2 and metal contacts, as shown in Figure 7a [84]. They pointed out that a thin tunnel layer with a sub−2 nm thickness could allow efficient tunneling, as shown in Figure 7b. The remarkably suppressed Fermi-level pinning has also been demonstrated with other dielectric layers, such as h−BN [82,85,86], ZnO [87] and TiO_2 [14]. Kwon et al. reported that defect-free vdW contacts were formed via a metal-deposition process with a selenium buffer layer on 2D layers, as shown in Figure 7c [88]. The device obeyed the Schottky-Mott rule and had a Fermi-level pinning factor of 0.91. Andrews et al. achieved a low Schottky barrier height of 25 meV by using $MoSe_2$ as an interlayer between MoS_2 channel and Ti electrodes, as shown in Figure 7d [89]. The reduction in Schottky barrier height can be attributed to the synergetic effect of Fermi-level pinning close to the conduction band edge of the $MoSe_2$ interlayer and the favorable conduction band offset between the $MoSe_2$ interlayer and MoS_2 channel, as shown in Figure 7e,f.

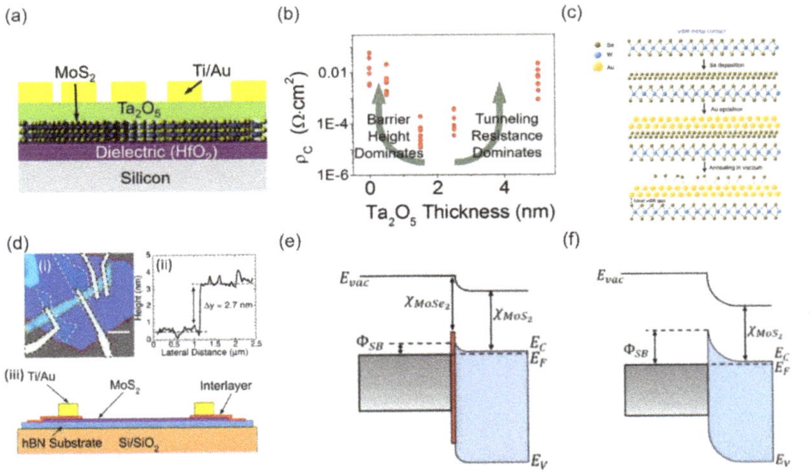

Figure 7. (**a**) Schematic of MoS_2 FETs with various thicknesses of Ta_2O_5 interlayer. (**b**) Measured specific contact resistivity as a function of Ta_2O_5 dielectric thickness. Reprinted/adapted with permission from Ref. [84]. Copyright 2016, American Chemical Society. (**c**) Schematic of vdW Au contact WSe_2 fabrication process. Reprinted/adapted with permission from Ref. [88]. Copyright 2022, Springer Nature. (**d**) (i) Optical micrograph of MoS_2 FETs with $MoSe_2$ interlayers. (ii) Thickness analysis of the $MoSe_2$ interlayers. (iii) Device structure of MoS_2 FETs with Ti/$MoSe_2$ interlayer contacts. Illustrations of the band alignments at the contacts with a $MoSe_2$ interlayer (**e**) and direct metal contacts (**f**). Reprinted/adapted with permission from Ref. [89]. Copyright 2020, American Chemical Society.

4. Determination Methods of Contact Resistance

The determination method of contact resistance in 2D FETs should be consistent in different works for ease of comparison. There are three commonly used methods, including the transfer length method (TLM), Y−function method and four-point probe method.

4.1. Transfer Length Method

The transfer length method is widely used in 2D FETs to extract contact resistance [35]. The device should be fabricated with different channel lengths, as shown in Figure 8a. R_{total} is the resistance between sourse and drain electrodes, R_{sh} is the channel sheet resistance and W is the channel width. When R_{total}/W is plotted versus the channel length, the y-axis intercept of the fitting line is equal to $2R_c$.

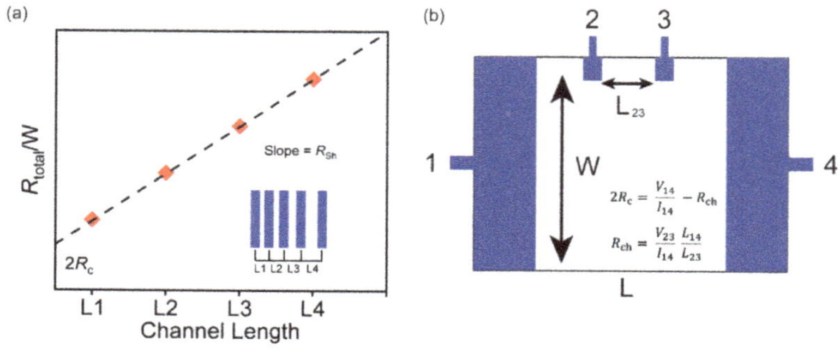

Figure 8. Methods to extract contact resistance. (**a**) Schematic of transfer length method. (**b**) Schematic of four-point probe method.

The contact resistance value extracted by TLM can have large variation when the channel length is large and the sheet resistance is huge. To minimize the estimated error, short-channel devices should be used, and statistic results are preferred.

4.2. Y-Function Method

The Y-function method requires only one transfer curve $I_d - I_g$ at the linear regime by applying a large gate voltage and a small source-drain voltage $V_d \ll V_g$ [90]. When the transconductance starts to decrease, the contact effect will dominate the μ attenuation and the contact resistance can be derived.

We assume that the contact resistance is comparable with the channel resistance. The source-drain voltage will drop at the contact region and I_d can be expressed as the following equation

$$I_d = \frac{\mu_0}{1+\theta_0(V_g - V_{th})} C_i \frac{W}{L}(V_g - V_{th} - 0.5V_d)(V_d - I_d R_c) \tag{4}$$

where μ_0, θ_0 and V_{th} are the intrinsic mobility in the linear regime, first-order mobility attenuation coefficient, and the threshold voltage, respectively. When $V_g - V_{th} \gg 0.5V_d$, $0.5V_d$ can be ignored. The effective mobility attenuation factor θ represents the contribution from both θ_0 and R_c. Therefore, I_d can be written as the following equation

$$I_d = \frac{\mu_0}{1+\theta(V_g - V_{th})} C_i \frac{W}{L}(V_g - V_{th}) V_d \tag{5}$$

The Y—function was defined as

$$Y = \frac{I_d}{\sqrt{g_m}} = \frac{I_d}{\sqrt{I_d/[1+\theta(V_g - V_{th})](V_g - V_{th})}} = \sqrt{\mu_0 C_i V_d \frac{W}{L}}(V_g - V_{th}) \tag{6}$$

where g_m is transconductance $g_m = \partial I_d / \partial V_g$. The value s_1 can be extracted from the slope of the Y−function versus V_g. The value s_2 can be extracted from the slope of $\frac{1}{\sqrt{g_m}}$ versus V_g. The R_c follows the equation:

$$R_c = \frac{s_2}{s_1} V_d \tag{7}$$

4.3. Four-Point Probe Method

The four-point probe method to extract contact resistance requires the fabrication of a device with the desired structure, as shown in Figure 8b. The contact resistance is given by the following equation:

$$2R_c = \frac{V_{14}}{I_{14}} - \frac{V_{23}}{I_{14}}\frac{L_{14}}{L_{23}} \tag{8}$$

5. Conclusions and Outlook

This review focused on the contact engineering of 2D FETs and discussed the origins of high contact resistance, the structure of top contact and edge contact, and the contact engineering in both structures. We believe that Fermi-level pinning in 2D devices is dominantly induced by interfacial gap states, and the solution to this challenge is to make a sharp and clean vdW interface at the contact regions. The top contact is compatible with the conventional Si CMOS process, but it is very challenging to control the deposition condition to achieve a vdW interface. The edge contact can be used in both top- and bottom—gate 2D FETs, but the accurate etching of 2D materials with little damage should be developed by using the dry etching technologies, such as reactive ion etching, plasma etching and inductively coupled plasma etching. Realization of the edge contact requires a much more complex fabrication process than that of the top contact. Although the edge contact methods can often achieve ultra-low contact resistance, the small contact area still limits the electrical performance of 2D FETs, such as on-state current. The insertion of a buffer layer provides another pathway to reduce the Fermi-level pinning effect, which can be combined with other contact-engineering strategies. In 2D GAAFETs, it is necessary to vertically integrate 2D FETs into integrated circuits, which is more challenging to achieve a good metal contact.

Overall, it is important to develop a CMOS-compatible contact deposition process to achieve large-scale 2D FETs with high-performance transport properties. One promising method is to employ an alloy composed of low-melting-point metal and high-melting point-metal as contact to simultaneously achieve vdW contact and increase the temperature endurance for the BEOL process. Another promising method is to build a high-quality mixed contact by combining the advantages of edge contact and top contact to overcome the small contact areas and Fermi-level pinning.

Author Contributions: J.M., X.Z., Y.T. and Y.Z. conceived the project. J.M. and Y.Z. wrote the manuscript. All authors have read and agreed to the published version of the manuscript.

Funding: We thank ZJU Micro-Nano Fabrication Center and ZJU-Hangzhou International Innovation Center for the supports. The project was primarily supported by the National Natural Science Foundation of China (62090030, 62090034, 62104214, 61974078); the Young Elite Scientists Sponsorship Program by CAST (2021QNRC001); the Opening Project of Key Laboratory of Optoelectronic Chemical Materials and Devices of Ministry of Education, Jianghan University (No. JDGD-202202); and Kun-Peng Program of Zhejiang Province.

Institutional Review Board Statement: Not applicable.

Informed Consent Statement: Not applicable.

Data Availability Statement: Not applicable.

Conflicts of Interest: The authors declare no conflict of interest.

References

1. Liu, Y.; Duan, X.; Shin, H.-J.; Park, S.; Huang, Y.; Duan, X. Promises and prospects of two-dimensional transistors. *Nature* **2021**, *591*, 43–53. [CrossRef] [PubMed]
2. Liu, L.; Li, T.; Ma, L.; Li, W.; Gao, S.; Sun, W.; Dong, R.; Zou, X.; Fan, D.; Shao, L.; et al. Uniform nucleation and epitaxy of bilayer molybdenum disulfide on sapphire. *Nature* **2022**, *605*, 69–75. [CrossRef] [PubMed]
3. Kim, Y.; Kim, T.; Lee, J.; Choi, Y.S.; Moon, J.; Park, S.Y.; Lee, T.H.; Park, H.K.; Lee, S.A.; Kwon, M.S.; et al. Tailored Graphene Micropatterns by Wafer-Scale Direct Transfer for Flexible Chemical Sensor Platform. *Adv. Mater.* **2021**, *33*, 2004827. [CrossRef]
4. Zhao, Y.; Xu, K.; Pan, F.; Zhou, C.; Zhou, F.; Chai, Y. Doping, Contact and Interface Engineering of Two-Dimensional Layered Transition Metal Dichalcogenides Transistors. *Adv. Funct. Mater.* **2017**, *27*, 1603484. [CrossRef]
5. Yu, Z.; Ning, H.; Cheng, C.C.; Li, W.; Liu, L.; Meng, W.; Luo, Z.; Li, T.; Cai, S.; Wang, P.; et al. Reliability of Ultrathin High−κ Dielectrics on Chemical-Vapor Deposited 2D Semiconductors. In Proceedings of the 2020 IEEE International Electron Devices Meeting (IEDM), San Francisco, CA, USA, 12–18 December 2020; pp. 3.2.1–3.3.4.
6. Sebastian, A.; Pendurthi, R.; Choudhury, T.H.; Redwing, J.M.; Das, S. Benchmarking monolayer MoS_2 and WS_2 field-effect transistors. *Nat. Commun.* **2021**, *12*, 693. [CrossRef]

7. Shen, P.-C.; Su, C.; Lin, Y.; Chou, A.-S.; Cheng, C.-C.; Park, J.-H.; Chiu, M.-H.; Lu, A.-Y.; Tang, H.-L.; Tavakoli, M.M.; et al. Ultralow contact resistance between semimetal and monolayer semiconductors. *Nature* **2021**, *593*, 211–217. [CrossRef]
8. Wang, Y.; Kim, J.C.; Li, Y.; Ma, K.Y.; Hong, S.; Kim, M.; Shin, H.S.; Jeong, H.Y.; Chhowalla, M. P-type electrical contacts for two-dimensional transition metal dichalcogenides. *Nature* **2022**, *610*, 61–66. [CrossRef]
9. Wu, R.; Tao, Q.; Li, J.; Li, W.; Chen, Y.; Lu, Z.; Shu, Z.; Zhao, B.; Ma, H.; Zhang, Z.; et al. Bilayer tungsten diselenide transistors with on-state currents exceeding 1.5 milliamperes per micrometre. *Nat. Electron.* **2022**, *5*, 497–504. [CrossRef]
10. Badaroglu, M. More Moore. In Proceedings of the 2021 IEEE International Roadmap for Devices and Systems Outbriefs, Santa Clara, CA, USA, 30 November 2021; pp. 1–38.
11. Schottky, W. Zur Halbleitertheorie der Sperrschicht- und Spitzengleichrichter. *Z. Für Phys.* **1939**, *113*, 367–414. [CrossRef]
12. Mott, N.F. The theory of crystal rectifiers. *Proc. Royal Soc. A Math. Phys. Eng. Sci.* **1939**, *171*, 27–38. [CrossRef]
13. Liu, Y.; Guo, J.; Zhu, E.; Liao, L.; Lee, S.-J.; Ding, M.; Shakir, I.; Gambin, V.; Huang, Y.; Duan, X. Approaching the Schottky–Mott limit in van der Waals metal–semiconductor junctions. *Nature* **2018**, *557*, 696–700. [CrossRef] [PubMed]
14. Kim, G.-S.; Kim, S.-H.; Park, J.; Han, K.H.; Kim, J.; Yu, H.-Y. Schottky Barrier Height Engineering for Electrical Contacts of Multilayered MoS_2 Transistors with Reduction of Metal-Induced Gap States. *ACS Nano* **2018**, *12*, 6292–6300. [CrossRef] [PubMed]
15. Tersoff, J. Schottky Barrier Heights and the Continuum of Gap States. *Phys. Rev. Lett.* **1984**, *52*, 465–468. [CrossRef]
16. Guo, Y.; Liu, D.; Robertson, J. 3D Behavior of Schottky Barriers of 2D Transition-Metal Dichalcogenides. *ACS Appl. Mater. Interfaces* **2015**, *7*, 25709–25715. [CrossRef]
17. Szczęśniak, D.; Hoehn, R.D.; Kais, S. Canonical Schottky barrier heights of transition metal dichalcogenide monolayers in contact with a metal. *Phys. Rev. B* **2018**, *97*, 195315. [CrossRef]
18. Shautsova, V.; Sinha, S.; Hou, L.; Zhang, Q.; Tweedie, M.; Lu, Y.; Sheng, Y.; Porter, B.F.; Bhaskaran, H.; Warner, J.H. Direct Laser Patterning and Phase Transformation of 2D $PdSe_2$ Films for On-Demand Device Fabrication. *ACS Nano* **2019**, *13*, 14162–14171. [CrossRef]
19. Choi, M.S.; Lee, M.; Ngo, T.D.; Hone, J.; Yoo, W.J. Chemical Dopant-Free Doping by Annealing and Electron Beam Irradiation on 2D Materials. *Adv. Electron. Mater.* **2021**, *7*, 2100449. [CrossRef]
20. Li, F.; Gao, F.; Xu, M.; Liu, X.; Zhang, X.; Wu, H.; Qi, J. Tuning Transport and Photoelectric Performance of Monolayer MoS_2 Device by E-Beam Irradiation. *Adv. Mater. Interfaces* **2018**, *5*, 1800348. [CrossRef]
21. Zhao, Y.; Bertolazzi, S.; Maglione, M.S.; Rovira, C.; Mas-Torrent, M.; Samorì, P. Molecular Approach to Electrochemically Switchable Monolayer MoS_2 Transistors. *Adv. Mater.* **2020**, *32*, 2000740. [CrossRef]
22. Matsunaga, M.; Higuchi, A.; He, G.; Yamada, T.; Krüger, P.; Ochiai, Y.; Gong, Y.; Vajtai, R.; Ajayan, P.M.; Bird, J.P.; et al. Nanoscale-Barrier Formation Induced by Low−Dose Electron-Beam Exposure in Ultrathin MoS_2 Transistors. *ACS Nano* **2016**, *10*, 9730–9737. [CrossRef]
23. Poddar, P.K.; Zhong, Y.; Mannix, A.J.; Mujid, F.; Yu, J.; Liang, C.; Kang, J.-H.; Lee, M.; Xie, S.; Park, J. Resist-Free Lithography for Monolayer Transition Metal Dichalcogenides. *Nano Lett.* **2022**, *22*, 726–732. [CrossRef] [PubMed]
24. Wu, R.J.; Udyavara, S.; Ma, R.; Wang, Y.; Chhowalla, M.; Birol, T.; Koester, S.J.; Neurock, M.; Mkhoyan, K.A. Visualizing the metal−MoS_2 contacts in two-dimensional field-effect transistors with atomic resolution. *Phys. Rev. Mater.* **2019**, *3*, 111001. [CrossRef]
25. Smyth, C.M.; Addou, R.; Hinkle, C.L.; Wallace, R.M. Origins of Fermi-level pinning between Molybdenum Dichalcogenides ($MoSe_2$, $MoTe_2$) and Bulk Metal Contacts: Interface Chemistry and Band Alignment. *J. Phys. Chem. C* **2019**, *123*, 23919–23930. [CrossRef]
26. Das, S.; Chen, H.-Y.; Penumatcha, A.V.; Appenzeller, J. High Performance Multilayer MoS_2 Transistors with Scandium Contacts. *Nano Lett.* **2013**, *13*, 100–105. [CrossRef]
27. Kim, C.; Moon, I.; Lee, D.; Choi, M.S.; Ahmed, F.; Nam, S.; Cho, Y.; Shin, H.-J.; Park, S.; Yoo, W.J. Fermi Level Pinning at Electrical Metal Contacts of Monolayer Molybdenum Dichalcogenides. *ACS Nano* **2017**, *11*, 1588–1596. [CrossRef]
28. Yuan, H.; Cheng, G.; Yu, S.; Hight Walker, A.R.; Richter, C.A.; Pan, M.; Li, Q. Field effects of current crowding in metal−MoS_2 contacts. *Appl. Phys. Lett.* **2016**, *108*, 103505. [CrossRef]
29. Taur, Y.; Ning, T.H. *Fundamentals of Modern VLSI Devices*; Cambridge University Press: Cambridge, UK, 2021.
30. Kwon, J.; Lee, J.-Y.; Yu, Y.-J.; Lee, C.-H.; Cui, X.; Hone, J.; Lee, G.-H. Thickness-dependent Schottky barrier height of MoS_2 field-effect transistors. *Nanoscale* **2017**, *9*, 6151–6157. [CrossRef]
31. Wang, J.; Guo, X.; Yu, Z.; Ma, Z.; Liu, Y.; Chan, M.; Zhu, Y.; Wang, X.; Chai, Y. Steep Slope p-type 2D WSe_2 Field-Effect Transistors with Van Der Waals Contact and Negative Capacitance. In Proceedings of the 2018 IEEE International Electron Devices Meeting (IEDM), San Francisco, CA, USA, 1–5 December 2018; pp. 22.3.1–22.3.4. [CrossRef]
32. Han, B.; Zhao, Y.; Ma, C.; Wang, C.; Tian, X.; Wang, Y.; Hu, W.; Samorì, P. Asymmetric Chemical Functionalization of Top-Contact Electrodes: Tuning the Charge Injection for High-Performance MoS_2 Field-Effect Transistors and Schottky Diodes. *Adv. Mater.* **2022**, *34*, 2109445. [CrossRef]
33. Mao, J.-Y.; Wu, S.; Ding, G.; Wang, Z.-P.; Qian, F.-S.; Yang, J.-Q.; Zhou, Y.; Han, S.-T. A van der Waals Integrated Damage-Free Memristor Based on Layered 2D Hexagonal Boron Nitride. *Small* **2022**, *18*, 2106253. [CrossRef]
34. Jung, Y.; Choi, M.S.; Nipane, A.; Borah, A.; Kim, B.; Zangiabadi, A.; Taniguchi, T.; Watanabe, K.; Yoo, W.J.; Hone, J.; et al. Transferred via contacts as a platform for ideal two-dimensional transistors. *Nat. Electron.* **2019**, *2*, 187–194. [CrossRef]

35. English, C.D.; Shine, G.; Dorgan, V.E.; Saraswat, K.C.; Pop, E. Improved Contacts to MoS_2 Transistors by Ultra-High Vacuum Metal Deposition. *Nano Lett.* **2016**, *16*, 3824–3830. [CrossRef] [PubMed]
36. Wang, Y.; Kim, J.C.; Wu, R.J.; Martinez, J.; Song, X.; Yang, J.; Zhao, F.; Mkhoyan, A.; Jeong, H.Y.; Chhowalla, M. Van der Waals contacts between three-dimensional metals and two-dimensional semiconductors. *Nature* **2019**, *568*, 70–74. [CrossRef] [PubMed]
37. Kumar, A.; Schauble, K.; Neilson, K.M.; Tang, A.; Ramesh, P.; Wong, H.S.P.; Pop, E.; Saraswat, K. Sub−200 Ω·μm Alloyed Contacts to Synthetic Monolayer MoS_2. In Proceedings of the 2021 IEEE International Electron Devices Meeting (IEDM), San Francisco, CA, USA, 11–16 December 2021; pp. 7.3.1–7.3.4. [CrossRef]
38. Jin, L.; Koester, S.J. High-Performance Dual-Gated Single-Layer WS_2 MOSFETs With Bi Contacts. *IEEE Electron Device Lett.* **2022**, *43*, 639–642. [CrossRef]
39. Chou, A.S.; Wu, T.; Cheng, C.C.; Zhan, S.S.; Ni, I.C.; Wang, S.Y.; Chang, Y.C.; Liew, S.L.; Chen, E.; Chang, W.H.; et al. Antimony Semimetal Contact with Enhanced Thermal Stability for High Performance 2D Electronics. In Proceedings of the 2021 IEEE International Electron Devices Meeting (IEDM), San Francisco, CA, USA, 11–16 December 2021; pp. 7.2.1–7.2.4. [CrossRef]
40. Liu, Y.; Stradins, P.; Wei, S.-H. Van der Waals metal-semiconductor junction: Weak Fermi level pinning enables effective tuning of Schottky barrier. *Sci. Adv.* **2016**, *2*, e1600069. [CrossRef] [PubMed]
41. Murali, K.; Dandu, M.; Watanabe, K.; Taniguchi, T.; Majumdar, K. Accurate Extraction of Schottky Barrier Height and Universality of Fermi Level De-Pinning of van der Waals Contacts. *Adv. Funct. Mater.* **2021**, *31*, 2010513. [CrossRef]
42. Zhang, C.; Gong, C.; Nie, Y.; Min, K.-A.; Liang, C.; Oh, Y.J.; Zhang, H.; Wang, W.; Hong, S.; Colombo, L.; et al. Systematic study of electronic structure and band alignment of monolayer transition metal dichalcogenides in Van der Waals heterostructures. *2D Mater.* **2016**, *4*, 015026. [CrossRef]
43. Liu, Y.; Wu, H.; Cheng, H.-C.; Yang, S.; Zhu, E.; He, Q.; Ding, M.; Li, D.; Guo, J.; Weiss, N.O.; et al. Toward Barrier Free Contact to Molybdenum Disulfide Using Graphene Electrodes. *Nano Lett.* **2015**, *15*, 3030–3034. [CrossRef]
44. Chuang, H.-J.; Chamlagain, B.; Koehler, M.; Perera, M.M.; Yan, J.; Mandrus, D.; Tománek, D.; Zhou, Z. Low-Resistance 2D/2D Ohmic Contacts: A Universal Approach to High-Performance WSe_2, MoS_2, and $MoSe_2$ Transistors. *Nano Lett.* **2016**, *16*, 1896–1902. [CrossRef]
45. Jang, J.; Ra, H.-S.; Ahn, J.; Kim, T.W.; Song, S.H.; Park, S.; Taniguch, T.; Watanabe, K.; Lee, K.; Hwang, D.K. Fermi-level pinning-Free WSe_2 Transistors via 2D Van der Waals Metal Contacts and Their Circuits. *Adv. Mater.* **2022**, *34*, 2109899. [CrossRef]
46. Zhao, Y.; Qiao, J.; Yu, Z.; Yu, P.; Xu, K.; Lau, S.P.; Zhou, W.; Liu, Z.; Wang, X.; Ji, W.; et al. High-Electron-Mobility and Air-STable 2D Layered $PtSe_2$ FETs. *Adv. Mater.* **2017**, *29*, 1604230. [CrossRef]
47. Wang, Y.; Li, L.; Yao, W.; Song, S.; Sun, J.T.; Pan, J.; Ren, X.; Li, C.; Okunishi, E.; Wang, Y.-Q.; et al. Monolayer $PtSe_2$, a New Semiconducting Transition-Metal-Dichalcogenide, Epitaxially Grown by Direct Selenization of Pt. *Nano Lett.* **2015**, *15*, 4013–4018. [CrossRef] [PubMed]
48. Zhang, K.; Yan, M.; Zhang, H.; Huang, H.; Arita, M.; Sun, Z.; Duan, W.; Wu, Y.; Zhou, S. Experimental evidence for type−II Dirac semimetal in $PtSe_2$. *Phys. Rev. B* **2017**, *96*, 125102. [CrossRef]
49. Shi, J.; Huan, Y.; Hong, M.; Xu, R.; Yang, P.; Zhang, Z.; Zou, X.; Zhang, Y. Chemical Vapor Deposition Grown Large-Scale Atomically Thin Platinum Diselenide with Semimetal–Semiconductor Transition. *ACS Nano* **2019**, *13*, 8442–8451. [CrossRef]
50. Das, T.; Yang, E.; Seo, J.E.; Kim, J.H.; Park, E.; Kim, M.; Seo, D.; Kwak, J.Y.; Chang, J. Doping-Free All $PtSe_2$ Transistor via Thickness-Modulated Phase Transition. *ACS Appl. Mater. Interfaces* **2021**, *13*, 1861–1871. [CrossRef] [PubMed]
51. Zhang, X.; Yu, H.; Tang, W.; Wei, X.; Gao, L.; Hong, M.; Liao, Q.; Kang, Z.; Zhang, Z.; Zhang, Y. All-van-der-Waals Barrier-Free Contacts for High-Mobility Transistors. *Adv. Mater.* **2022**, *34*, 2109521. [CrossRef] [PubMed]
52. Mootheri, V.; Arutchelvan, G.; Banerjee, S.; Sutar, S.; Leonhardt, A.; Boulon, M.-E.; Huyghebaert, C.; Houssa, M.; Asselberghs, I.; Radu, I. Graphene based Van der Waals contacts on MoS_2 field effect transistors. *2D Mater.* **2021**, *8*, 015003. [CrossRef]
53. Choi, M.S.; Ali, N.; Ngo, T.D.; Choi, H.; Oh, B.; Yang, H.; Yoo, W.J. Recent Progress in 1D Contacts for 2D-Material-Based Devices. *Adv. Mater.* **2022**, *34*, 2202408. [CrossRef]
54. Liu, G.-B.; Xiao, D.; Yao, Y.; Xu, X.; Yao, W. Electronic structures and theoretical modelling of two-dimensional group-VIB transition metal dichalcogenides. *Chem. Soc. Rev.* **2015**, *44*, 2643–2663. [CrossRef]
55. Yang, Z.; Kim, C.; Lee, K.Y.; Lee, M.; Appalakondaiah, S.; Ra, C.-H.; Watanabe, K.; Taniguchi, T.; Cho, K.; Hwang, E.; et al. A Fermi-Level-Pinning-Free 1D Electrical Contact at the Intrinsic 2D MoS_2–Metal Junction. *Adv. Mater.* **2019**, *31*, 1808231. [CrossRef]
56. Jain, A.; Szabó, Á.; Parzefall, M.; Bonvin, E.; Taniguchi, T.; Watanabe, K.; Bharadwaj, P.; Luisier, M.; Novotny, L. One-Dimensional Edge Contacts to a Monolayer Semiconductor. *Nano Lett.* **2019**, *19*, 6914–6923. [CrossRef]
57. Mirabelli, G.; McGeough, C.; Schmidt, M.; McCarthy, E.K.; Monaghan, S.; Povey, I.M.; McCarthy, M.; Gity, F.; Nagle, R.; Hughes, G.; et al. Air sensitivity of MoS_2, $MoSe_2$, $MoTe_2$, HfS_2, and $HfSe_2$. *J. Appl. Phys.* **2016**, *120*, 125102. [CrossRef]
58. Illarionov, Y.Y.; Waltl, M.; Rzepa, G.; Kim, J.-S.; Kim, S.; Dodabalapur, A.; Akinwande, D.; Grasser, T. Long-Term Stability and Reliability of Black Phosphorus Field-Effect Transistors. *ACS Nano* **2016**, *10*, 9543–9549. [CrossRef] [PubMed]
59. Wang, Y.; Gao, J.; Wei, B.; Han, Y.; Wang, C.; Gao, Y.; Liu, H.; Han, L.; Zhang, Y. Reduction of the ambient effect in multilayer InSe transistors and a strategy toward sTable 2D-based optoelectronic applications. *Nanoscale* **2020**, *12*, 18356–18362. [CrossRef]
60. Chai, Y.; Ionescu, R.; Su, S.; Lake, R.; Ozkan, M.; Ozkan, C.S. Making one-dimensional electrical contacts to molybdenum disulfide-based heterostructures through plasma etching. *Phys. Status Solidi A* **2016**, *213*, 1358–1364. [CrossRef]
61. Cheng, Z.; Yu, Y.; Singh, S.; Price, K.; Noyce, S.G.; Lin, Y.-C.; Cao, L.; Franklin, A.D. Immunity to Contact Scaling in MoS_2 Transistors Using in Situ Edge Contacts. *Nano Lett.* **2019**, *19*, 5077–5085. [CrossRef] [PubMed]

62. Duerloo, K.-A.N.; Li, Y.; Reed, E.J. Structural phase transitions in two-dimensional Mo− and W−dichalcogenide monolayers. *Nat. Commun.* **2014**, *5*, 4214. [CrossRef]
63. Voiry, D.; Mohite, A.; Chhowalla, M. Phase engineering of transition metal dichalcogenides. *Chem. Soc. Rev.* **2015**, *44*, 2702–2712. [CrossRef]
64. Huang, H.H.; Fan, X.; Singh, D.J.; Zheng, W.T. Recent progress of TMD nanomaterials: Phase transitions and applications. *Nanoscale* **2020**, *12*, 1247–1268. [CrossRef]
65. Kim, J.; Park, H.; Yoo, S.; Im, Y.-H.; Kang, K.; Kim, J. Defect-Engineered n-Doping of WSe_2 via Argon Plasma Treatment and Its Application in Field-Effect Transistors. *Adv. Mater. Interfaces* **2021**, *8*, 2100718. [CrossRef]
66. Ma, R.; Zhang, H.; Yoo, Y.; Degregorio, Z.P.; Jin, L.; Golani, P.; Ghasemi Azadani, J.; Low, T.; Johns, J.E.; Bendersky, L.A.; et al. $MoTe_2$ Lateral Homojunction Field-Effect Transistors Fabricated using Flux-Controlled Phase Engineering. *ACS Nano* **2019**, *13*, 8035–8046. [CrossRef]
67. Lee, E.K.; Abdullah, H.; Torricelli, F.; Lee, D.H.; Ko, J.K.; Kim, H.H.; Yoo, H.; Oh, J.H. Boosting the Optoelectronic Properties of Molybdenum Diselenide by Combining Phase Transition Engineering with Organic Cationic Dye Doping. *ACS Nano* **2021**, *15*, 17769–17779. [CrossRef]
68. Kappera, R.; Voiry, D.; Yalcin, S.E.; Branch, B.; Gupta, G.; Mohite, A.D.; Chhowalla, M. Phase-engineered low-resistance contacts for ultrathin MoS_2 transistors. *Nat. Mater.* **2014**, *13*, 1128–1134. [CrossRef] [PubMed]
69. Zhao, Y.; Bertolazzi, S.; Samorì, P. A Universal Approach toward Light-Responsive Two-Dimensional Electronics: Chemically Tailored Hybrid van der Waals Heterostructures. *ACS Nano* **2019**, *13*, 4814–4825. [CrossRef] [PubMed]
70. Cho, S.; Kim, S.; Kim, J.H.; Zhao, J.; Seok, J.; Keum, D.H.; Baik, J.; Choe, D.-H.; Chang, K.J.; Suenaga, K.; et al. Phase patterning for ohmic homojunction contact in $MoTe_2$. *Science* **2015**, *349*, 625–628. [CrossRef] [PubMed]
71. Keum, D.H.; Cho, S.; Kim, J.H.; Choe, D.-H.; Sung, H.-J.; Kan, M.; Kang, H.; Hwang, J.-Y.; Kim, S.W.; Yang, H.; et al. Bandgap opening in few-layered monoclinic $MoTe_2$. *Nat. Phys.* **2015**, *11*, 482–486. [CrossRef]
72. Naz, M.; Hallam, T.; Berner, N.C.; McEvoy, N.; Gatensby, R.; McManus, J.B.; Akhter, Z.; Duesberg, G.S. A New 2H−2H′/1T Cophase in Polycrystalline MoS_2 and $MoSe_2$ Thin Films. *ACS Appl. Mater. Interfaces* **2016**, *8*, 31442–31448. [CrossRef] [PubMed]
73. Ma, Y.; Liu, B.; Zhang, A.; Chen, L.; Fathi, M.; Shen, C.; Abbas, A.N.; Ge, M.; Mecklenburg, M.; Zhou, C. Reversible Semiconducting-to-Metallic Phase Transition in Chemical Vapor Deposition Grown Monolayer WSe_2 and Applications for Devices. *ACS Nano* **2015**, *9*, 7383–7391. [CrossRef]
74. Zhu, J.; Wang, Z.; Yu, H.; Li, N.; Zhang, J.; Meng, J.; Liao, M.; Zhao, J.; Lu, X.; Du, L.; et al. Argon Plasma Induced Phase Transition in Monolayer MoS_2. *J. Am. Chem. Soc.* **2017**, *139*, 10216–10219. [CrossRef]
75. Oyedele, A.D.; Yang, S.; Feng, T.; Haglund, A.V.; Gu, Y.; Puretzky, A.A.; Briggs, D.; Rouleau, C.M.; Chisholm, M.F.; Unocic, R.R.; et al. Defect-Mediated Phase Transformation in Anisotropic Two-Dimensional $PdSe_2$ Crystals for Seamless Electrical Contacts. *J. Am. Chem. Soc.* **2019**, *141*, 8928–8936. [CrossRef]
76. Cai, X.; Wu, Z.; Han, X.; Chen, Y.; Xu, S.; Lin, J.; Han, T.; He, P.; Feng, X.; An, L.; et al. Bridging the gap between atomically thin semiconductors and metal leads. *Nat. Commun.* **2022**, *13*, 1777. [CrossRef]
77. Li, H.; Cheng, M.; Wang, P.; Du, R.; Song, L.; He, J.; Shi, J. Reducing Contact Resistance and Boosting Device Performance of Monolayer MoS_2 by In Situ Fe Doping. *Adv. Mater.* **2022**, *34*, 2200885. [CrossRef] [PubMed]
78. Vu, V.T.; Vu, T.T.H.; Phan, T.L.; Kang, W.T.; Kim, Y.R.; Tran, M.D.; Nguyen, H.T.T.; Lee, Y.H.; Yu, W.J. One-Step Synthesis of $NbSe_2$/Nb−Doped−WSe_2 Metal/Doped−Semiconductor van der Waals Heterostructures for Doping Controlled Ohmic Contact. *ACS Nano* **2021**, *15*, 13031–13040. [CrossRef] [PubMed]
79. Hemanjaneyulu, K.; Kumar, J.; Shrivastava, M. MoS_2 Doping Using Potassium Iodide for Reliable Contacts and Efficient FET Operation. *IEEE Trans. Electron Devices* **2019**, *66*, 3224–3228. [CrossRef]
80. Khan, M.F.; Rehman, S.; Akhtar, I.; Aftab, S.; Ajmal, H.M.S.; Khan, W.; Kim, D.-K.; Eom, J. High mobility $ReSe_2$ field effect transistors: Schottky-barrier-height-dependent photoresponsivity and broadband light detection with Co decoration. *2D Mater.* **2020**, *7*, 015010. [CrossRef]
81. Sarkar, D.; Xie, X.; Kang, J.; Zhang, H.; Liu, W.; Navarrete, J.; Moskovits, M.; Banerjee, K. Functionalization of Transition Metal Dichalcogenides with Metallic Nanoparticles: Implications for Doping and Gas-Sensing. *Nano Lett.* **2015**, *15*, 2852–2862. [CrossRef]
82. Wang, J.; Yao, Q.; Huang, C.-W.; Zou, X.; Liao, L.; Chen, S.; Fan, Z.; Zhang, K.; Wu, W.; Xiao, X.; et al. High Mobility MoS_2 Transistor with Low Schottky Barrier Contact by Using Atomic Thick h−BN as a Tunneling Layer. *Adv. Mater.* **2016**, *28*, 8302–8308. [CrossRef]
83. Chen, J.-R.; Odenthal, P.M.; Swartz, A.G.; Floyd, G.C.; Wen, H.; Luo, K.Y.; Kawakami, R.K. Control of Schottky Barriers in Single Layer MoS_2 Transistors with Ferromagnetic Contacts. *Nano Lett.* **2013**, *13*, 3106–3110. [CrossRef]
84. Lee, S.; Tang, A.; Aloni, S.; Philip Wong, H.S. Statistical Study on the Schottky Barrier Reduction of Tunneling Contacts to CVD Synthesized MoS_2. *Nano Lett.* **2016**, *16*, 276–281. [CrossRef]
85. Mleczko, M.J.; Yu, A.C.; Smyth, C.M.; Chen, V.; Shin, Y.C.; Chatterjee, S.; Tsai, Y.-C.; Nishi, Y.; Wallace, R.M.; Pop, E. Contact Engineering High-Performance n-Type $MoTe_2$ Transistors. *Nano Lett.* **2019**, *19*, 6352–6362. [CrossRef]
86. Cui, X.; Shih, E.-M.; Jauregui, L.A.; Chae, S.H.; Kim, Y.D.; Li, B.; Seo, D.; Pistunova, K.; Yin, J.; Park, J.-H.; et al. Low-Temperature Ohmic Contact to Monolayer MoS_2 by van der Waals Bonded Co/h−BN Electrodes. *Nano Lett.* **2017**, *17*, 4781–4786. [CrossRef]

87. Jang, J.; Kim, Y.; Chee, S.-S.; Kim, H.; Whang, D.; Kim, G.-H.; Yun, S.J. Clean Interface Contact Using a ZnO Interlayer for Low-Contact-Resistance MoS_2 Transistors. *ACS Appl. Mater. Interfaces* **2020**, *12*, 5031–5039. [CrossRef] [PubMed]
88. Kwon, G.; Choi, Y.-H.; Lee, H.; Kim, H.-S.; Jeong, J.; Jeong, K.; Baik, M.; Kwon, H.; Ahn, J.; Lee, E.; et al. Interaction- and defect-free van der Waals contacts between metals and two-dimensional semiconductors. *Nat. Electron.* **2022**, *5*, 241–247. [CrossRef]
89. Andrews, K.; Bowman, A.; Rijal, U.; Chen, P.-Y.; Zhou, Z. Improved Contacts and Device Performance in MoS_2 Transistors Using a 2D Semiconductor Interlayer. *ACS Nano* **2020**, *14*, 6232–6241. [CrossRef]
90. Chang, H.-Y.; Zhu, W.; Akinwande, D. On the mobility and contact resistance evaluation for transistors based on MoS_2 or two-dimensional semiconducting atomic crystals. *Appl. Phys. Lett.* **2014**, *104*, 113504. [CrossRef]

nanomaterials

MDPI

Review

Evolution Application of Two-Dimensional MoS_2-Based Field-Effect Transistors

Chunlan Wang [1,*], Yongle Song [1] and Hao Huang [2]

[1] School of Science, Xi'an Polytechnic University, Xi'an 710048, China
[2] Guangxi Key Laboratory of Processing for Nonferrous Metals and Featured Material, School of Resources, Environment and Materials, Guangxi University, Nanning 530004, China
* Correspondence: wangchunlan@whu.edu.cn

Abstract: High-performance and low-power field-effect transistors (FETs) are the basis of integrated circuit fields, which undoubtedly require researchers to find better film channel layer materials and improve device structure technology. MoS_2 has recently shown a special two-dimensional (2D) structure and superior photoelectric performance, and it has shown new potential for next-generation electronics. However, the natural atomic layer thickness and large specific surface area of MoS_2 make the contact interface and dielectric interface have a great influence on the performance of MoS_2 FET. Thus, we focus on its main performance improvement strategies, including optimizing the contact behavior, regulating the conductive channel, and rationalizing the dielectric layer. On this basis, we summarize the applications of 2D MoS_2 FETs in key and emerging fields, specifically involving logic, RF circuits, optoelectronic devices, biosensors, piezoelectric devices, and synaptic transistors. As a whole, we discuss the state-of-the-art, key merits, and limitations of each of these 2D MoS_2-based FET systems, and prospects in the future.

Keywords: MoS_2-FETs; logic and radio-frequency circuits; photodetector; biosensor; piezoelectric devices; synapses transistors

1. Introduction

The FET is voltage-adjustable electronic device that controls the output circuit current by controlling the electric field effect of the input circuit, which can work at very low current and voltage, and which can easily integrate on a piece of silicon and other substrates, so the FET has been widely used in large-scale integrated circuits [1]. The common FET device structure consists of a gate, source/drain, channel layer and dielectric. As is well known, the continuous miniaturization of silicon (Si)-based FETs has driven the exponential growth of integrated circuits for more than half a century. However, with the bulk thickness reduced to less than 10 nm, Si FETs exhibit a large number of attenuation phenomena of carrier mobility. This requires researchers to come up with a series of strategies to overcome this limitation, such as finding new materials, developing new structures, and improving processes [1,2].

Recently, 2D materials including graphene [3], black phosphorus (BP) [4], and transition metal dichalcogenides (TMDs) [5–8] have shown natural advantages for the further reduction in FET sizes due to their natural atomic-level thickness and surface without hanging bonds, for remedying the shortcomings of Si-based FETs. Table 1 summarizes the main electric performance of MoS_2-based FETs and other mainstream 2D materials-based FETs. The graphene FET has ultra-high mobility. However, its I_{on}/I_{off} ratio is lowest (typical below 10) owing to the lack of bandgap, and it is difficult to apply to logic electronics. The mobility of the BP FET is much higher than that of the MoS_2 FET, but it is not stable, due to reaction with water and oxygen in the air to decompose. TMDs (MoS_2, WS_2, WSe_2, etc.)-based FETs have an ultrahigh I_{on}/I_{off} ratio and good mobility, which can be assembled into low-power electronics. Especially, MoS_2 FETs have the advantages of higher mobility, higher I_{on}/I_{off} ratio, and lower subthreshold swing, compared with other TMD FETs. Overall, MoS_2 is the most suitable channel layer material for FETs, and MoS_2 FETs are the promising candidate for downscaling electronics with a short channel, low thickness, small volume, fast speed, high sensitivity, light weight, etc.

Table 1. Overview summary comparison of emerging 2D materials-based FETs.

Materials	Technology	Configuration	Contact Electrode	Dielectric	Mobility (cm^2/Vs)	I_{on}/I_{off}	SS (mV/dec)	Ref.
MoS_2	Exfoliation	Dual-gated	Ni/Au	Al_2O_3	517	10^8	140	[7]
	CVD	Back-gated	Ti/Au	HfO_2	118	10^8	/	[9]
	Au-assisted Exfoliation	Back-gated	Ti/Au	SiO_2	25	10^7	100	[10]
	MOCVD	Top-gated	Au/Ti	SiO_2	22	10^5	/	[11]
	VLS	Back-gated	Au	SiO_2	33	10^8	980	[12]
	PLD	Top-gated	Au/Ti	HfO_2	9	10^5	/	[13]
	Exfoliation	Top-gated	Au	SiO_2/Cytop	31	10^7	/	[14]
	Exfoliation	Back-gated	Graphene	SiO_2	9	10^6	/	[15]
	APCVD	Back-gated	Cr/Au	SiO_2	54	10^8	/	[16]
	Exfoliation	Top-gated	Cr/Au	PMMA/P(VDF-TrFE)	/	10^7	/	[17]
WSe_2	CVD	Back-gated	Ti/Pd	BN	92	10	/	[5]
$MoSe_2$	Exfoliation	Back-gated	Ni	SiO_2	50	10^6	/	[8]
WS_2	Exfoliation	Back-gated	Ti/Au	SiO_2	20	10^6	70	[6]
BP	Exfoliation	Top-gated	Ni/Au	SiO_2	862	10^2	563	[4]
Graphene	CVD	Back-gated	Cr/Au	TiO_2	1872	—	/	[3]

Table 1 also shows that different MoS_2 FETs have performed differently, because the natural atomic-level thickness and large specific surface area of MoS_2 make the interface quality have a great influence on the MoS_2 properties [2]. High-quality MoS_2 preparation methods have constantly been investigated in recent years. The common preparation methods of MoS_2 include Chemical Vapor Deposition (CVD), Metal–Organic Chemical Vapor Deposition (MOCVD), Atomic-layer deposition (ALD), and Vapor–Liquid–Solid (VLS) [18]. The properties of stripped MoS_2 are better than those of CVD-prepared MoS_2, because stripped MoS_2 has high quality and less impurities, while the CVD process will bring a lot of impurities. In addition, a 6-inch uniform monolayer MoS_2 can be grown by CVD in a short time [19]. MoS_2 with the low surface energy can also be easily stripped and transferred, the size of which depends mainly on the size of the bulk MoS_2, which can reach the centimeter level [20].

However, in MoS_2 FETs, a high-quality MoS_2 between the contact electrode and the dielectric layer will still be obtained at the contact interface and dielectric interface, which has further led to the different performance of MoS_2 FETs [21]. Therefore, for improving the performance of MoS_2 FETs, we mainly focus on optimizing the contact behavior [14], regulating the conductive channel [16], and rationalizing the dielectric layer [17]. This is also the subject of our review. Then, we discuss the applications of MoS_2 FETs in key and emerging fields [21], involving logic and RF circuits [22,23], optoelectronic devices [24], biosensors [25], piezoelectric devices [26], and synaptic transistors [27] (Figure 1).

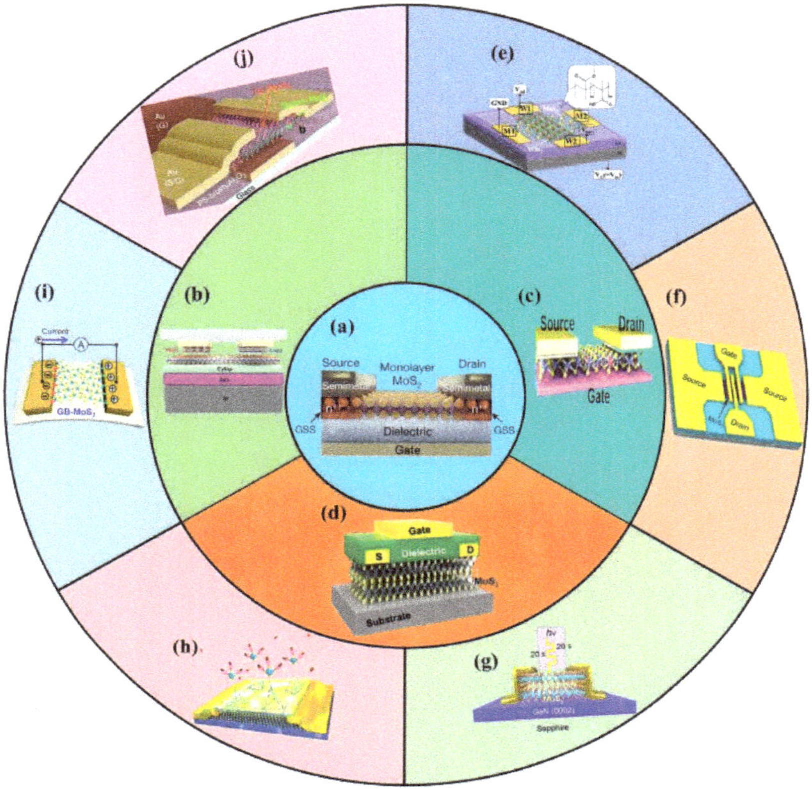

Figure 1. MoS_2 FET performance improvement strategies and applications. (**a**) Schematic of monolayer MoS_2 FET (reprinted/adapted with permission from Ref. [21]. Copyright 2021 Springer Nature). (**b**) Improving MoS_2 FET performance by optimizing contact behavior (reprinted/adapted with permission from Ref. [14]. Copyright 2022 Wiley). (**c**) Improving MoS_2 FET performance by regulating conductive channel (reprinted/adapted with permission from Ref. [16]. Copyright 2022 Wiley). (**d**) Improving MoS_2 FET performance by rationalizing dielectric layer (reprinted/adapted with permission from Ref. [17]. Copyright 2020 IEEE). The applications of MoS_2 FET in (**e**) logic circuits (reprinted/adapted with permission from Ref. [22]. Copyright 2020 American Chemical Society); (**f**) radio-frequency circuits (reprinted/adapted with permission from Ref. [23]. Copyright 2014 American Chemical Society); (**g**) optoelectronic devices (reprinted/adapted with permission from Ref. [24]. Copyright 2021 American Chemical Society); (**h**) biosensors (reprinted/adapted with permission from Ref. [25]. Copyright 2021 Elsevier); (**i**) piezoelectric devices (reprinted/adapted with permission from Ref. [26]. Copyright 2020 American Chemical Society); (**j**) synaptic transistors (reprinted/adapted with permission from Ref. [27]. Copyright 2022 American Chemical Society).

2. Performance Improvement Strategy of MoS_2 FETs

2.1. Optimizing Contact Behavior

A metal electrode and traditional 2D electrode are usually directly deposited by chemical or physical deposition methods on MoS_2 surfaces, which damages the contact interface, resulting in the Schottky barrier and Fermi-level pinning effect, between the MoS_2/metal electrode, and the MoS_2/traditional 2D semiconductor in MoS_2 FETs [28], which could reduce electrical performance [29]. Therefore, it is very important to find suitable electrode materials to form ohmic contact and eliminate the Fermi pinning effect [30]. Recently, the van der Waals contact has been an indirect method of preparation into MoS_2, which does not use direct chemical bonding, avoiding the damage of the deposition process and the diffusion at the contact interface, Here, we review the methods for obtaining ultra-low contact resistance in both 2D and non-2D contacts.

Shen et al. prepared a back-gated monolayer MoS_2 FET, where bismuth, nickel, and titanium were used as electrodes to explore the contact barrier. The results showed that the Bi-MoS_2 FET had the lowest contact barrier (123 Ω μm at a carrier density of 1.5×10^{13} cm^{-2}), attributed to the suppression of metal-induced interstitial states by the semi-metallic bismuth contact with MoS_2. This mechanism also contributes to good ohmic contact in other TMDs FETs, i.e., the I_{on}/I_{off} ratios are as high as (10^7) at low voltages (1.5 V) [21].

Samori et al. prepared an asymmetric Schottky diode by self-assembled monolayers (SAMs) of pre-functionalized gold electrodes, then transferring them to MoS_2 by the dry method [14]. Figure 2a shows the MoS_2 FETs structure, where the drain/source electrodes are 2, 3, 4, 5, 6-pentafluorobenzenethiol (PFBT)-functionalized electrodes and the source 4-(dimethylamine) benzenethiol (DABT) functionalized electrode, respectively. The output curve of the Schottky diode is shown in Figure 2b. The V_{gs} of the device is reduced from 90 to −30 V, and the rectification ratio reaches a maximum of 10^3 at −30 V in Figure 2c. Figure 2d–f show the band structure of devices under different bias voltages. This strategy can be adjusted to control the electrode and different chemisorption SAMs functionalization to reduce (increase) the charge injection barrier, thus providing a strategy for the manufacture of asymmetric charge injection devices [14].

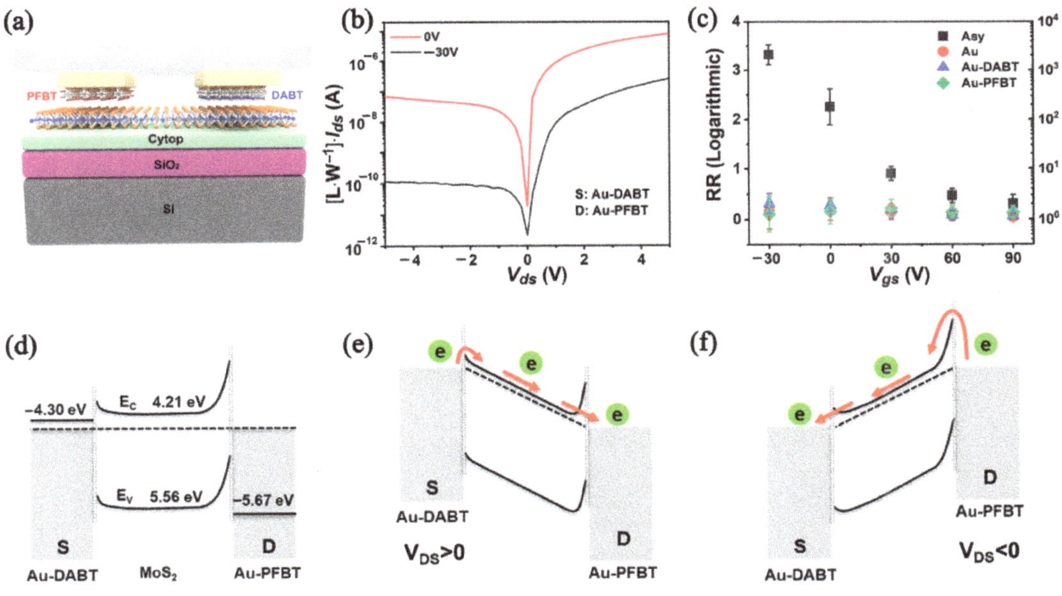

Figure 2. (**a**) Structure diagram of asymmetric MoS_2 FETs. (**b**) Semilogarithmic plot of the I_{ds}-V_{ds}. (**c**) Rectification ratio. (**d**–**f**) Energy band diagram at thermodynamic equilibrium, positively biased source–drain, and negatively biased source–drain, respectively. (Reprinted/adapted with permission from Ref. [14]. Copyright 2022 Wiley).

Recently, experiments have found that graphene materials as the electrodes of MoS_2 FETs can also achieve good ohmic characteristics [31]. Yu et al. fabricated graphene/MoS_2 heterojunction FETs, by adjusting the Fermi level of the graphene electrode to modulate the height of the Schottky barrier between heterojunctions, where the short-channel effect with a channel length less than 30 nm was successfully eliminated, and at the same time, the drain-induced barrier lowering of the device was 0.92 v/v and the I_{on}/I_{off} ratio was as high as 10^8 [31]. Kim et al. reported high properties of monolayer MoS_2 FETs with a nitrogen-doped graphene (NGr) electrode [15], as shown in Figure 3. Compared to undoped graphene electrodes, the device current increased 214% and the field-effect mobility increased fourfold. Figure 3c,f illustrates that the Fermi level of MoS_2 FET was improved by employing the NGr electrode. Therefore, this is one of the effective means to reduce contact resistance, which will provide a new idea for the development of high-performance devices.

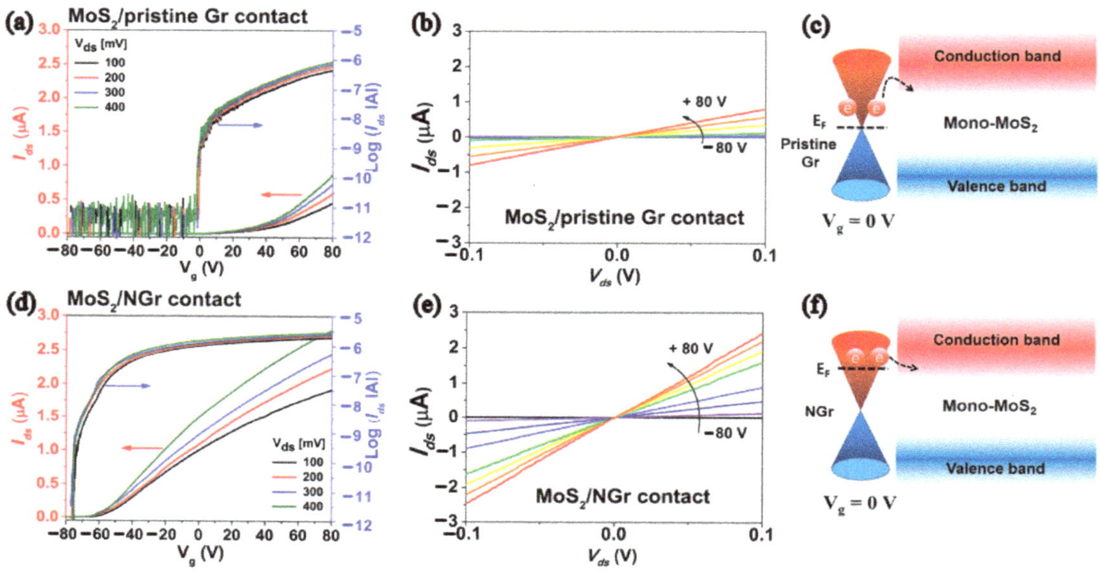

Figure 3. Electrical characteristics of MoS_2/pristine Gr contact device: (**a**) Transfer characteristic (left axis) and semilogarithm (right axis). (**b**) Output characteristic. (**c**) Schematic diagram of band alignment. Electrical characteristics of MoS_2/NGr contact device: (**d**) Transfer characteristic (left axis) and semilogarithm (right axis). (**e**) Output characteristic. (**f**) Schematic diagram of band alignment. (Reprinted/adapted with permission from Ref. [15]. Copyright 2019 American Institute of Physics).

Two-dimensional Mxenes contain transition metal carbides, carbonitrides, and nitrides, and have graphene-like surface structures and good electrical conductivity [32], which show great application potential in barriers, capacitors, and electrodes of electrochemical systems [33]. Dai et al. calculated the contact barrier of MoS_2 FETs with Ta_2C, Ta_2CF_2, and $Ta_2C(OH)_2$ as the electrode material, using density functional theory. The results showed that the N-type Schottky barrier was created by a Ta_2C electrode. However, using a Ta_2CF_2 or $Ta_2C(OH)_2$ electrode can form ohmic contact, and the resistance of $MoS_2/Ta_2C(OH)_2$ was 2 times smaller than that of MoS_2/Ta_2CF_2. This study provides theoretical guidance for the application of MXene materials in MoS_2 FETs [30]. Du et al. first studied the contact characteristics between MoS_2 and $M_3C_2(OH)_2$ (M = Ti, Zr, Hf) by first principles, which found that Ti and Hf are more suitable for ohmic contact as electrode materials. The subthreshold swing (SS) range of devices was 100~200 mV/decade, and the I_{on}/I_{off} ratio was as high as 10^6, in the sub-10 nm range [34].

To sum up, in the modification of the MoS_2 FET electrode, the selection of an appropriate metal electrode or other excellent electrode materials can reduce the Schottky barrier, especially the combination of the above new two-dimensional material, and MoS_2 can better achieve a high-performance, low-power-consumption FET, which will also be a significant attraction for the development of electronic devices in the future. In addition, 2D MoS_2 could produce unique quantum effects by contacting with the metal, such as quantum transport, superconductors, and valley transport [35].

2.2. Boosting Conductive Channel

MoS_2 with the good direct bandgap (~1.8 eV) when the bulk MoS_2 is stripped to monolayer MoS_2 sets the stage for MoS_2 FETs with high mobility and high I_{on}/I_{off} ratio [36]. As is well known, doping is one of the most prevalent techniques to regulate the band structure of semiconductor materials [37], which have both metal ion doping and nonmetal ion doping. An oxygen uniformly doped monolayer MoS_2 can be prepared directly by in situ chemical vapor deposition on a 2-inch sapphire substrate. The results showed that the bandgap of MoS_2 was regulated (from 2.25 eV (intrinsic) to 1.72 eV (heavily doped). The mobility of the MoS_2-XOx FET was 78 cm^2/Vs, and the I_{on}/I_{off} ratio was 3.5×10^8 [38]. Shi et al. developed a one-step CVD method to achieve the growth of centimeter-level monolayer MoS_2 film [16]. Unidirectionally Fe-doped MoS_2 domains (domain size up to 250 μm) were prepared on 2-inch commercial c-plane sapphire, which achieved very low contact resistance (≈678 Ω μm) and good ohmic contact with the electrode, as shown in Figure 4a–c. MoS_2 FETs obtained electron mobility (54 $cm^2V^{-1}s^{-1}$ at room temperature, 94 cm^2/Vs at 100 K, and I_{on}/I_{off} ratio (10^8)). The electron mobility of monolayer Fe-MoS_2 was decreased with increasing temperature, as observed in Figure 4d,e, which indicated that it can inhibit the scattering of impurities. Figure 4f shows the energy barriers of pristine MoS_2 and Fe-MoS_2.

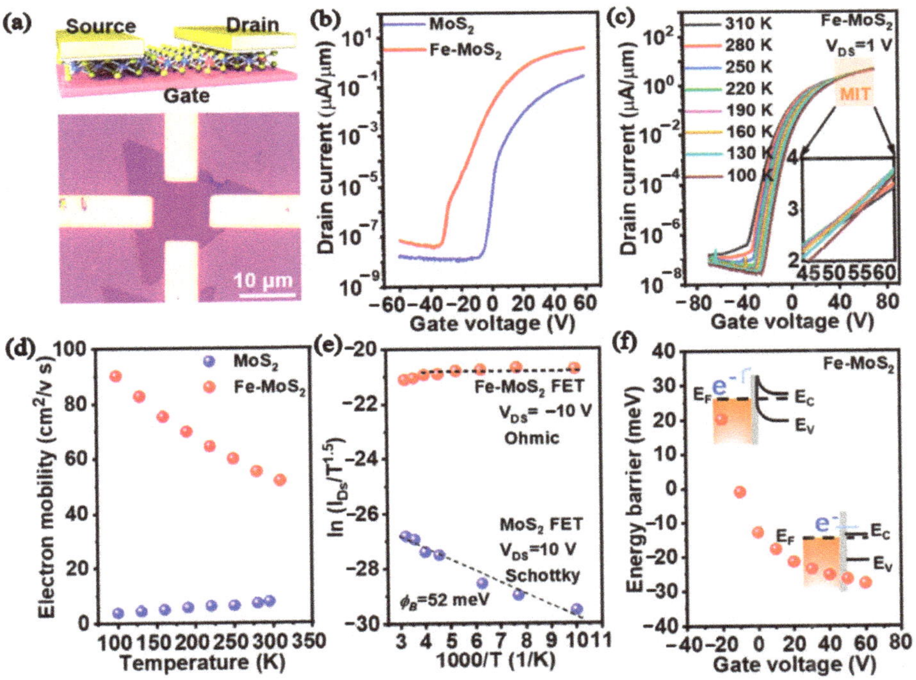

Figure 4. (**a**) Schematic diagram and OM graph of monolayer Fe-MoS_2 device. (**b**) Transfer characteristic of MoS_2-based devices. (**c**) Temperature vs. transfer characteristic. The inset shows the MIT area. (**d**) Electron mobility vs. temperature. (**e**) Arrhenius plots. (**f**) Energy barriers and energy band of the monolayer Fe-MoS_2. (Reprinted/adapted with permission from Ref. [16]. Copyright 2022 Wiley).

Wang et al. first synthesized a Ta-doped p-type monolayer MoS_2 by NaCl-assisted CVD, which has the advantages of large area, controllability, high quality, and controllable doping concentration [39]. With the addition of Ta, MoS_2 FET devices showed bipolar properties and changed from N type to P type with increasing concentration. The p-type MoS_2 active layer was applied with heavy niobium doping by mechanical stripping on the Si substrate. Then, the p-type MoS_2 FETs with the Pt electrode obtained a 0.13 eV contact barrier, output current of -10 μA, and drain voltage of -1 V when the channel length was ~1 μm [40]. Han et al. reported a strategy for the controllable transformation of n-type MoS_2 into p-type MoS_2 by low energy (100 eV) He^+ irradiation, as shown in Figure 5. Through theoretical calculation and characterization, it was found that this method increases the band size through the migration of the topmost S atom, and electron capture transforms n-type MoS_2 into p-type MoS_2 [41].

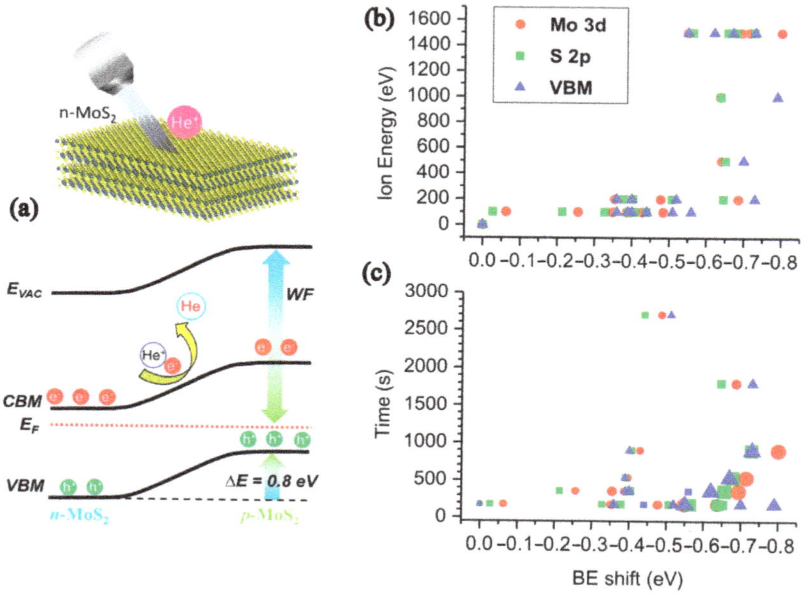

Figure 5. (**a**) Schematic diagram of the conversion of n-type MoS_2. The negative binding energy shifts of Mo 3d, S 2p, and valence band photoemission spectra depend on (**b**) the ion energy band, and (**c**) irradiation time of He+ ion irradiation. (Reprinted/adapted with permission from Ref. [41]. Copyright 2022 Springer Nature).

Doping can effectively improve the performance of MoS_2 films, but in the case of thin atom thickness or few layers, traditional doping strategies such as ion implantation or high-temperature diffusion could not improve the performance, because these methods tend to destroy the lattice and produce defects. Therefore, it is particularly important to find new doping methods, which provides inspiration for future research and development of more processes suitable for less layer MoS_2 doping.

2.3. Rationalizing Dielectric Layer

As for the MoS_2 FET, the surface accumulation of charges and charge traps on the dielectric are big issues leading to high leakage current, which is observed by using a scanning tunneling microscope [42,43]. Therefore, an appropriate dielectric layer with high permittivity should be used to effectively reduce the influence of impurity scattering at the interface on carrier transport in the channel layer, while avoiding the possible short-channel effect and reducing gate leakage current [44,45]. It is noted that a high-K dielectric layer (HfO_2, ZrO_2, HfZrO, etc.) can effectively shield the scattering of charged impurities and boost the gate control ability on the channel carriers, thus effectively improving the electrical performance of MoS_2 FETs [46–49]. However, the high-K dielectric layer will also generate trap charges and surface optical phonons to offset its advantages [49].

To solve these problems, Song et al. found that adding Al to the ZrO_2 dielectric layer in a certain proportion (Zr:Al = 1:1) could effectively improve the electrical performance of MoS_2 FETs, due to the moderate Al reducing the oxygen vacancy, and optimize leakage current, as well as improve the interface state density [50]. Zhao et al. demonstrated HfO_2 with treatment by NH_3 plasma and added Al_2O_3 as a high-K dielectric layer of the top-gate MoS_2 FETs. That significantly reduced the leakage current and provided a high carrier mobility ~87 cm^2/Vs, I_{on}/I_{off} ~2.1×10^7, and SS of 72 mV/dec [51]. Li et al. reported MoS_2 FETs with the self-limiting epitaxy technology, employing a monolayer molecular crystal of perylene-tetracarboxylic dianhydride as a buffer layer, graphene as a gate, and HfO_2 as a gate dielectric layer, which obtained good dielectric properties, low leakage current, and high breakdown power. The device has SS ~73 mV/dec, $I_{on}/I_{off} > 10^7$, and is not affected by short-channel effects [52]. Liao et al. designed high-performance MoS_2 FETs with the double-layer gate dielectric structure, as shown in Figure 6a,b. A high-K vinylidene fluoride-trifluoroethylene (P(VDF-TrFE)) dielectric is used to provide high carrier concentrations, and a low-K polymethyl methacrylate (PMMA) dielectric provides device stability, as shown in Figure 6d,e. MoS_2 FETs with low operating voltage, no hysteresis, and high stability were achieved through combining high-K and low-K dielectrics, as shown in Figure 6f [17].

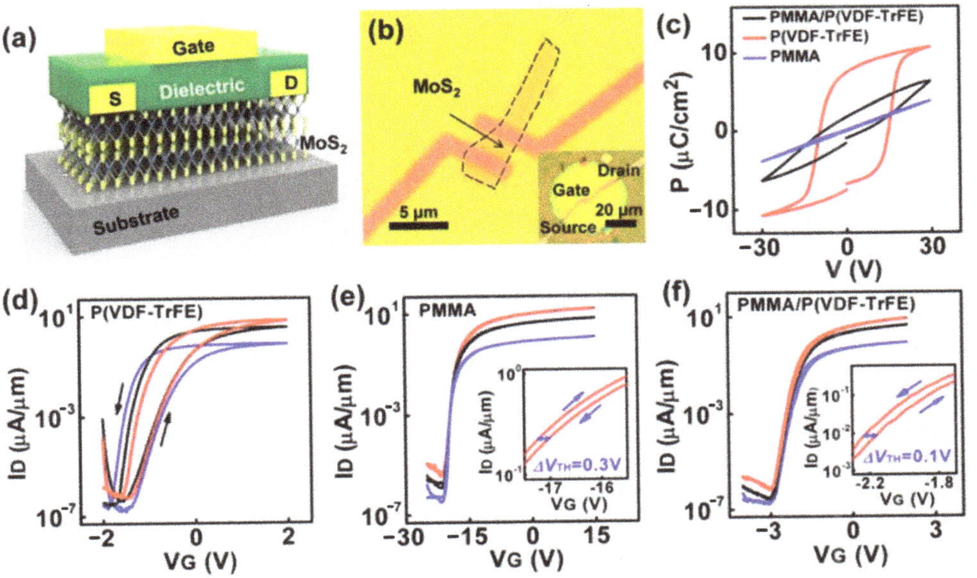

Figure 6. (**a**) Structure diagram of MoS_2 transistor. (**b**) The OM graph of MoS_2 transistor. (**c**) Ferroelectric characteristics of the different dielectrics. (**d–f**) Transfer properties of the MoS_2 transistors with different gate dielectrics, at V_D = 1 V (Red), 0.5 V (black), and 0.1 V (blue), receptively. The insets are the enlarged figure of the transfer curve with V_D = 1 V. (Reprinted/adapted with permission from Ref. [17]. Copyright 2020 IEEE).

Therefore, an excellent dielectric layer is important to obtain nonhysteresis, low voltage, and stable operation for MoS_2 FETs. A high-K dielectric can effectively suppress the scattering of charge impurities and the other problems. However, when the 2D MoS_2 material as the channel layer is very thin, the interface state problem will be more significant. Therefore, finding suitable single-layer/multi-layer dielectric materials and their high-quality deposition methods, or adding possible buffer layers, etc., can play a role in reducing these negative effects of monolayer MoS_2 FETs.

3. Logic and Radio-Frequency Circuits

3.1. Logic Circuits

Logical computation is an important part of a computer. Silicon complementary metal oxide semiconductor (CMOS) circuits are widely used at present, and they have complementary features to ensure data transmission will not produce problems, such as loss threshold voltage, accurate logic transmission, high I_{on}/I_{off}, and easy integration. The application of graphene in logic devices is limited by the zero bandgap. Therefore, the MoS_2 with a suitable bandgap provides a basis for the future application in logic computing devices [53]. Simulations revealed that monolayer MoS_2 FETs showed a 52% smaller drain-induced barrier lowering, and a 13% smaller SS, than the 3 nm thick-body Si FETs at a channel length of 10 nm [54]. Therefore, MoS_2 FETs are expected to improve the performance, and obtain new functions of devices in the field of electronics and display technology [55].

For voltage switching and high-frequency operation, logic devices need to achieve some performance, such as a high I_{on}/I_{off} ratio (>10^3) and moderate mobility [18]. Ang et al. prepared a top-gate MoS_2 FETs. The monolayer MoS_2 was grown on sapphire substrates by CVD and employed HfO_2 as a high-K dielectric layer. The inverter was fabricated by direct-coupled FET logic technology to obtain a high voltage gain ~16. When V_{DD} = 3 V, the total noise margin was 0.72 V_{DD} [56]. Pan et al. proposed a double-gate MoS_2 FETs to overcome the difficulty in regulating the threshold voltage and SS under the condition of a single gate. The device obtained an ultralow SS value of 65.5 mV/dec in the large current range above 10^4, when the dual gate was operating simultaneously in the inverter [57].

Multiple inverters require FETs with a stable resistance ratio to ensure constant output voltage. Kim et al. prepared cross-type FETs and the WSe_2/MoS_2 PN heterojunction surface was treated with PMMA-co-PMAA, so the channel current and anti-bipolar transistor region characteristics improved. Figure 7c,d shows that PMMA-co-PMAA can increase the current and ensure the stability of the inverter by doping effect-inducing charge transfer. Applying a cross-type p−n heterojunction WSe_2/MoS_2-based FET into a ternary inverter, three stable logic states of 1, 1/2, and 0 were realized [22], as shown in Figure 7a,b.

Zhou et al. fabricated MoS_2 FETs with a double-gate structure and double surface channel, where the top gate and the back gate serve as two input signals. Logic (OR, AND) was successfully implemented in a single cell [58]. MoS_2-based FETs will replace the traditional silicon device to achieve a higher degree of integration of logic devices, but only simple logic can be achieved at present. On the one hand, the development of P-type MoS_2 FETs still has a low performance and complex process problems, so its application in CMOS circuits still has a long way to go. On the other hand, mass production of large-size monolayer MoS_2 is also a difficulty. However, we still believe that MoS_2-based FETs will be applied to complex and scalable large-scale integrated circuits in the future.

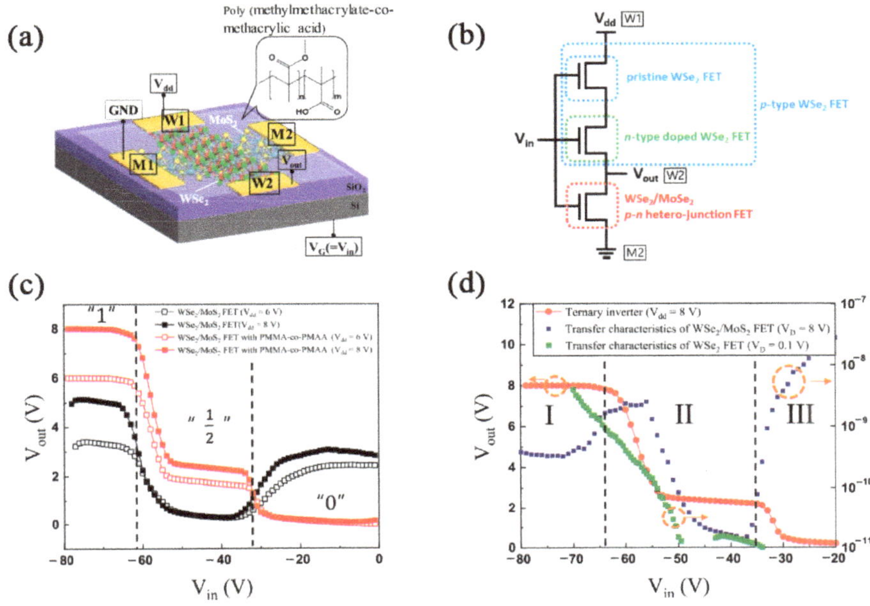

Figure 7. (**a**) Schematic diagram of ternary inverter by cross-type p–n heterojunction WSe_2/MoS_2-based FETs. (**b**) Logic circuit for ternary inverter. (**c**) V_{out} vs. V_{in} characteristic curves. (**d**) Characteristics of ternary inverter (red markers). Transfer characteristic curves of device (blue markers) and WSe_2-based FETs (green markers). (Reprinted/adapted with permission from Ref. [22]. Copyright 2020 American Chemical Society).

3.2. Radio-Frequency Circuits

With the advent of the 5G communication era, the position of RF devices is increasingly prominent and comparable to the importance of logic devices and memory devices. Facing the problem of bandwidth growth, promoting the working frequency of RF transistors is inevitable [53]. The maximum frequency of oscillation (f_{max}) based on III-V materials has exceeded 1 THz [59]. To further improve the operating frequency and bandwidth, optimizing the device processes or finding materials with higher mobility such as graphene is an effective strategy. Although the RF devices made of graphene can reach the operating frequency of mainstream RF devices, the gain cannot be further improved, due to the restriction of the zero bandgap. However, MoS_2 with a sizable bandgap can solve this bottleneck effectively [60].

Cheng et al. demonstrated a high-performance MoS_2 RF device on a flexible substrate with an intrinsic cut-off frequency of 42 GHz and f_{max} up to 50 GHz, and an intrinsic gain over 30 [61]. Wu et al. reported a microwave circuit with a MoS_2 self-switching diode. Ten layers of MoS_2 were grown on an Al_2O_3/Si substrate to realize the audio spectrum of an amplitude-modulated microwave signal in the 0.9~10 GHz band [62]. Modreanu et al. fabricated RF transistors with a double-layer MoS_2 channel structure, and the RF performance of the device was improved by electrostatic doping of the back gate. For MoS_2 RF transistors with a 190 nm gate length, at V_{BG} = 3 V, the external natural cutoff frequency was 6 GHz, the internal natural cutoff frequency was up to 19 GHz, and the maximum oscillation frequency was 29.7 GHz [63].

Among the flexible electronic devices, the f_T and f_{max} of MoS_2 RF transistors were higher than or comparable to those based on flexible electronic materials such as silicon film and InGaZnO [59]. Gao et al. transferred the bilayer MoS_2 onto a flexible polyimide substrate to prepare high-performance RF transistors. The different gate lengths (0.3 μm,

0.6 μm, and 1 μm) affected the performance of MoS_2 RF transistors. It was found that f_T and f_{max} increased with the decrease in the gate length, and with the gate length of 0.3 μm, the external f_T was 4 GHz and f_{Max} was 10 GHz [64].

Dresselhaus et al. formed a (1T/1T′-2H) phase heterostructure of MoS_2, and a Schottky diode flexible rectifier was fabricated by ohmic contact with palladium and Au electrodes, as shown in Figure 8d. Figure 8a–c show that the rectifier voltage increased with the increase in RF power, the maximum power efficiency could reach 40.1% at the 2.4 GHz band, and f_T was 10 MHz. A radio-frequency energy collector was prepared by integrating with a flexible Wi-Fi antenna, as shown in Figure 8e. The radio-frequency energy collector at 2.5 cm produced an output voltage of 250 mV at 5.9 GHz. These provided a universal energy-harvesting building block, which could also be integrated with various flexible electronic systems [65].

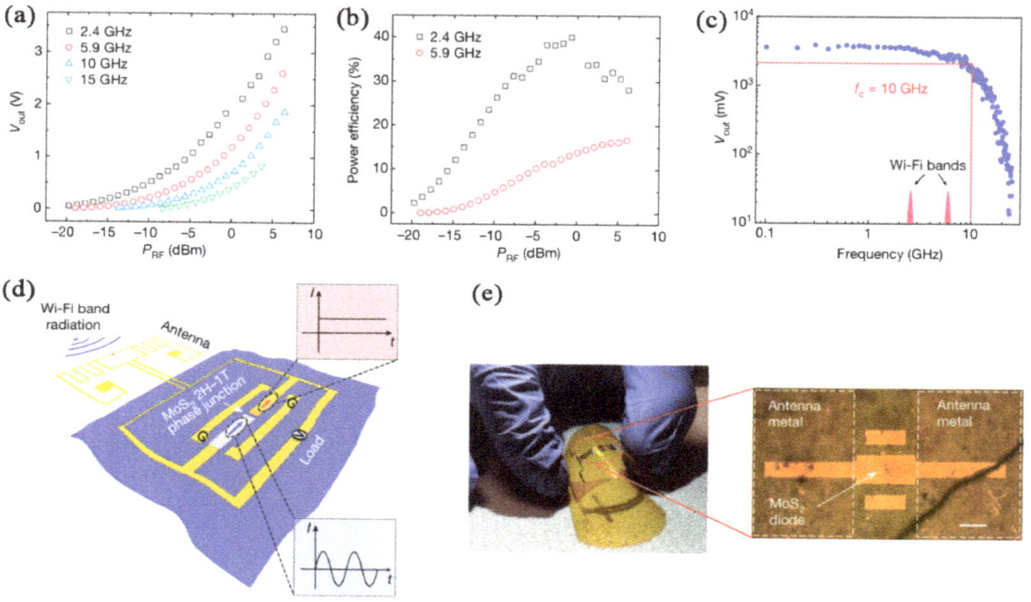

Figure 8. MoS_2 phase-junction rectenna as a wireless RF energy harvester: (**a**) Output voltage vs. input RF power. (**b**) Power efficiency of MoS_2 rectifiers vs. input power. (**c**) Output voltage vs. frequency. (**d**) Flexible MoS_2 wireless energy harvesting. The illustrative I–t curves correspond to the a.c. (blue arrow) and d.c. (red arrow) currents. (**e**) MoS_2 rectenna on Kapton. (Reprinted/adapted with permission from Ref. [65]. Copyright 2019 Springer Nature).

Most MoS_2-based RF transistors have been concentrated in bilayer structures because bilayer MoS_2 typically has higher mobility and saturation speed than single-layer MoS_2, providing a better power gain and cutoff frequency for the device. However, the high-frequency performance of MoS_2 RF transistors is still lower than that of silicon transistors. Therefore, it is important to optimize the device structure or process (such as optimizing gate structure and edge contact). In addition, MoS_2 with excellent mechanical properties can be used to manufacture flexible devices and, combined with RF, logic, and other fields, can achieve flexible electronic systems such as wearable devices, flexible sensors, and medical equipment.

4. Photodetectors

A photodetector is a device that converts photons into current by the photogenerated voltage effect of semiconductor materials. When a light source with strong radiation energy illuminates a semiconductor, the semiconductor material will absorb photons and generate electron–hole pairs to generate photocurrent [66]. MoS_2 has a wide bandgap, layered structure, strong photoluminescence characteristics, and excellent mechanical properties. Photodetectors made of MoS_2 have strong light response characteristics in the terahertz, mid-infrared, visible light, and near-infrared ranges [67]. Widely used photodetector performance indicators include responsivity, time response, specific detection rate, and spectral selectivity [68].

Zhang et al. developed a method for rapid synthesis of multilayer MoS_2 films from 1 L to more than 20 L with good quality using NaCl as a promoter. Then, a monolayer-to-multilayer MoS_2 photodetector was constructed, as shown in Figure 9a. Figure 9b,c illustrates the output curve of the monolayer–multilayer MoS_2 heterojunction device and the outstanding rectifying ratio (10^3). Figure 9d depicts the band diagram of the monolayer MoS_2 under high performance. In order to explore photoresponsivity, the transfer characteristic curve (Figure 9e) of the device under light and dark conditions and the light response ability under different voltages were investigated. Figure 9f shows that the device could achieve a maximum sensitivity of 10^4 A/W at 0 V (V_{GS}) [69].

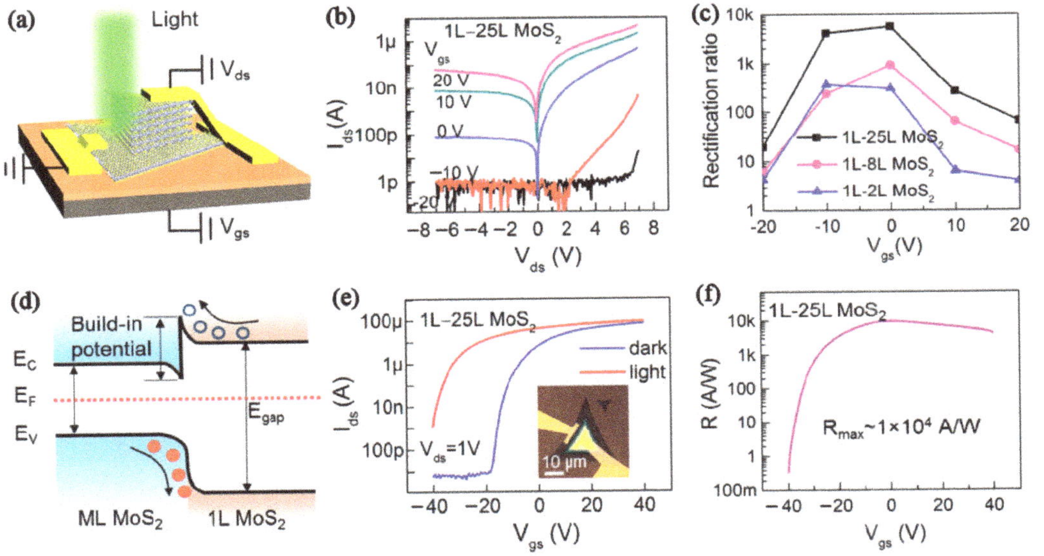

Figure 9. (**a**) Schematic of the monolayer–multilayer (1L–25L) MoS_2 heterojunction device. (**b**) I_{ds}-V_{ds} characteristics. (**c**) Rectification ratio as V_{gs}. (**d**) Band diagram of the 1 L-ML MoS_2 heterojunction in the off state ($V_{gs} = 0$; $V_{ds} = 0$). (**e**) I_{ds}-V_{gs} characteristics. The inset is the OM graph of the device. (**f**) V_{gs} vs. photoresponsivity. (Reprinted/adapted with permission from Ref. [69]. Copyright 2019 American Chemical Society).

Wu et al. prepared a photodetector by combining MXene nanoparticles with a 2D MoS_2 in a hybrid plasma structure. This strategy could improve the optical response of MoS_2 and make it more sensitive to visible light. The experimental results showed that the response rate of the device was 20.67 A/W, the detectivity was 5.39×10^{12} Jones, and the external quantum efficiency was up to 5167% [70]. Wang et al. grew a gate dielectric heap (Al_2O_3/HZO/TiN) on a Si substrate and transferred multilayer MoS_2 to prepare an ultra-sensitive negative-capacitance MoS_2 phototransistor. The HZO film could significantly enhance the optical gating effect, suppress dark current, and improve the ratio of light to dark current through a ferroelectric local electrostatic field and the ferroelectric NC effect. The experiment demonstrated that the prototype device had a high detection rate of 4.7×10^{14} cm $Hz^{1/2}W^{-1}$ and a high response rate of 96.8 AW^{-1} at low operating voltages of V_{ds} = 0.5 V and V_g = 1.6 V at room temperature [71]. Walila et al. grew a 3.5 μm GaN on a sapphire substrate, and the MoS_2 was mechanically stripped as the active layer. Figure 10a,b show the highly sensitive photodetector with gold and chromium as the electrode. Compared with bare GaN photodetectors, the responsivity and EQE of the photodetectors improved by 5 times, as shown in Figure 10d,e. Under the condition of 1 V bias and 365 nm excitation at 1 mW/cm^2, the highest response rate and EQE were obtained, which were 1.8×10^4 A/W and 6.19×10^6%, respectively [24]. The GaN/MoS_2 heterojunction laid a foundation for wide bandgaps and excellent photoelectric performance.

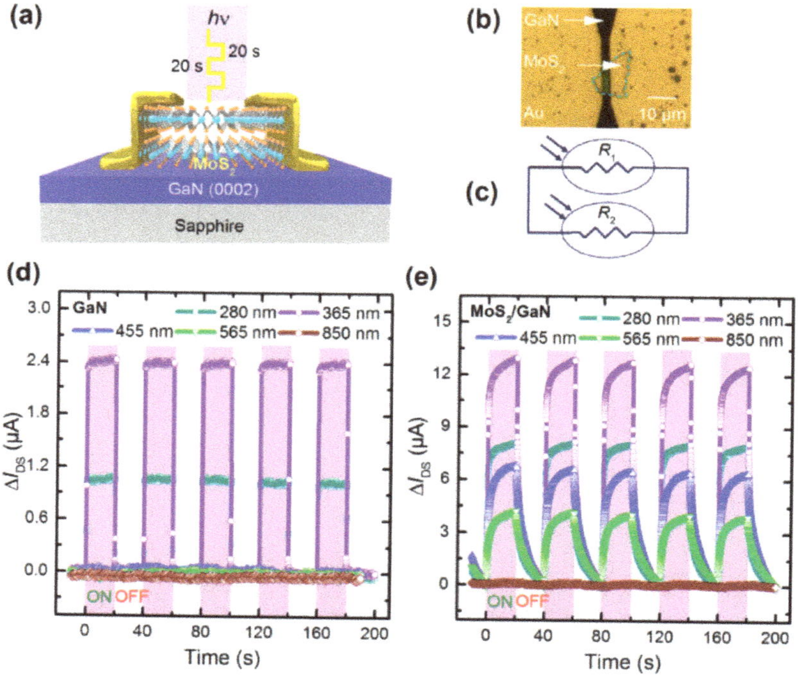

Figure 10. (**a**) Device structure of MoS_2/GaN photodetector. (**b**) OM graph of the hybrid device. (**c**) Equivalent circuit diagram of the hybrid device. Photoresponse measurements of (**d**) GaN devices and (**e**) MoS_2/GaN devices. (Reprinted/adapted with permission from Ref. [24]. Copyright 2021 American Chemical Society).

At present, the MoS_2 photodetector mainly focuses on two aspects: the change in layer number and the construction of a heterojunction. The monolayer MoS_2 has a low absorption surface and low quantum yield, which limits its development in photoelectric devices. In contrast, the multilayer MoS_2 has a small bandgap and can improve the absorption efficiency. Especially, the MoS_2-based heterojunction photodetector could achieve faster charge transfer and optical response. Optoelectronic devices require trade-offs between responsiveness and response time to meet practical needs. Although MoS_2-based photodetectors have some difficulties, they still have great advantages in flexible photodetectors and integrated nano-optoelectronic systems.

5. Biosensors

Biosensors are widely used in clinical and disease treatment as a powerful tool to detect biochemical processes. The great demand for detection has promoted the development of new nanomaterials as a sensing platform [72,73]. The global spread of COVID-19 has warned people to protect their own health, and biosensors with rapid, real-time, and accurate detection can effectively contain the spread of the virus and remind people to treat themselves in a timely manner [73,74]. MoS_2 attracted new interest with the multidimensional structures and structure-dependent unique electronic, electrocatalytic, and optical properties [75]. Therefore, MoS_2 has been widely applied in biosensors that can detect DNA, proteins, metal ions, and other compounds [76].

Arshad et al. modified Au nanoparticles on MoS_2 nanosheets to prepare bottom-gate FETs for the detection of the low-concentration C-reactive protein. The detection limit and sensitivity of BG-FETs were 8.38 fg/mL and 176 nA/g·mL^{-1}, respectively [77]. Dai et al. prepared MoS_2 on a 300 nm SiO_2/Si substrate and functionalized MoS_2 by combining five different DNA sequences into a DNA tetrahedron, as shown in Figure 11a. Figure 11b shows that the biosensor is extremely sensitive to the target protein (prostate-specific antigen, PSA). In phosphate-buffered brine, the detection limit was 1 fg/mL and the linear range was 1 fg/mL~100 ng/mL. Figure 11c shows that the I_{ds} decreases as the concentration increases, and bovine serum albumin (BSA) is used as an interference signal for comparison of detection effects. Figure 11d shows that the biosensor is extremely sensitive to the target protein (prostate-specific antigen, PSA) [25].

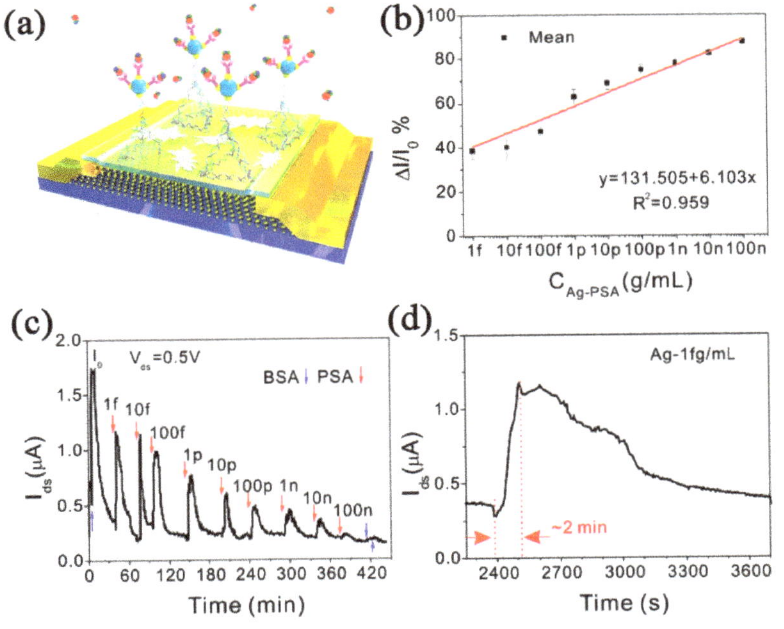

Figure 11. (**a**) Schematic diagram of 3D MoS_2 biosensor. (**b**) Real-time detection (V_{ds}= 0.5 V) of different concentrations. (**c**) Response time curve. (**d**) Response variation vs. PSA concentrations. (Reprinted/adapted with permission from Ref. [25]. Copyright 2021 Elsevier).

He et al. prepared a surface plasmon resonance (SPR) biosensor with Au nanoparticles-modified MoS_2 nanoflowers for IgG detection. The flower-like structure can provide more active sites for metal particles to react with target substances. The sensitivity of the MoS_2-Aunps-modified sensor was 0.0472 nm/(μg/mL), which is about 3 times higher than that of an unmodified sensor (0.016 nm/(μg/mL)). The limit of detection of IgG was reduced 2.7 times (from 0.16 to 0.06 μg/mL) [78]. Kim et al. prepared MoS_2 biological FETs with a nanoporous structure. Nanoporous structures could increase the edge area, using block copolymer photolithography. The surface area of the nanocycle was selectively functionalized by the newly formed suspended groups at the edge of the nanocycle. The biosensor exhibited superior detection performance in human serum and artificial saliva and resulted in a limit of detection of 1 ag/mL for cortisol [79].

This section mainly introduces the MoS_2-based biosensor and the biocompatibility of the MoS_2 material with biological cells, and the portability, sensitivity, and low power consumption of FETs have been widely studied and applied in biological monitoring. During the background of COVID-19, the development of wearable biosensors for rapid monitoring has become an urgent need.

6. Piezoelectric Devices

Piezoelectric properties exist in materials with centrosymmetric fractures. When strain was applied, the center of gravity of the cation and anion did not coincide, which resulted in a voltage potential at the interface between the semiconductor and the metal [80]. The mechanical flexibility and piezoelectric and photoelectric effects of MoS_2 materials can meet the needs of pressure sensors, micro-electro-mechanical systems, and active flexible electronic devices [81]. A layered MoS_2 material could be modified to obtain piezoelectric properties and could be applied to piezoelectric nanogenerators [82]. The piezoelectric effect can be used to collect micro-mechanical energy and convert it into electrical energy [83].

Hone et al. first reported the piezoelectric properties of monolayer MoS_2 in 2014 [84]. Kim et al. prepared a monolayer MoS_2 piezoelectric nanogenerator (PNG) by sulfur vacancy passivation. The output peak current and voltage of the PNG monolayer MoS_2 nanoflakes treated by S increased by 3 times (100 pA) and 2 times (22 mV), respectively. In addition, the maximum power increased by nearly 10 times [85]. Hu et al. fabricated a single-layer butterfly MoS_2 piezoelectric device on a polyethylene terephthalate substrate. Figure 12a,d shows the piezoelectric properties of the MoS_2 single crystal (SC-MoS_2) and MoS_2 with grain boundaries (GB-MoS_2) under external strain. Figure 12b,c shows that under the action of external stress, the current value generated by the direction of "armchair" was higher than that of the direction of "Zigzag". Figure 12e,f shows that the GB-MoS_2 current density is higher than that of SC-MoS_2. It was found that the piezoelectric effect induced by the grain boundary (~ 50%) could be applied to a self-powered sensor to monitor changes in human blood pressure [26].

Xue et al. reported a new self-powered NH_3 sensor, which employed monolayer MoS_2 materials on PET and covered polydimethylsiloxane films, and deposited an Au electrode. The sensor could be worn on different parts of the body and was responsive with a fast response/recovery time of 18 s/16 s [86]. Willatzen et al. first calculated the piezoelectric coefficient of 3R-MoS_2. Here, the 5-layer 3R-MoS_2 structure had the highest piezoelectric constant in all MoS_2 multilayer structures. The maximum piezoelectric constant was about 13% higher than that of the monolayer MoS_2 structure [87].

The discovery of piezoelectric effects could utilize a lot of neglected energy. At present, the application of MoS_2 piezoelectric properties is not mature enough. It is necessary to systematically study the piezoelectric properties, piezoelectric coefficients, and deformation direction of MoS_2 with different layers to help develop high-performance piezoelectric devices. Hence, MoS_2 piezoelectric sensors will realize their great potential in nanoscale electromechanical systems, micro-flexible wearable devices, and other fields in the future.

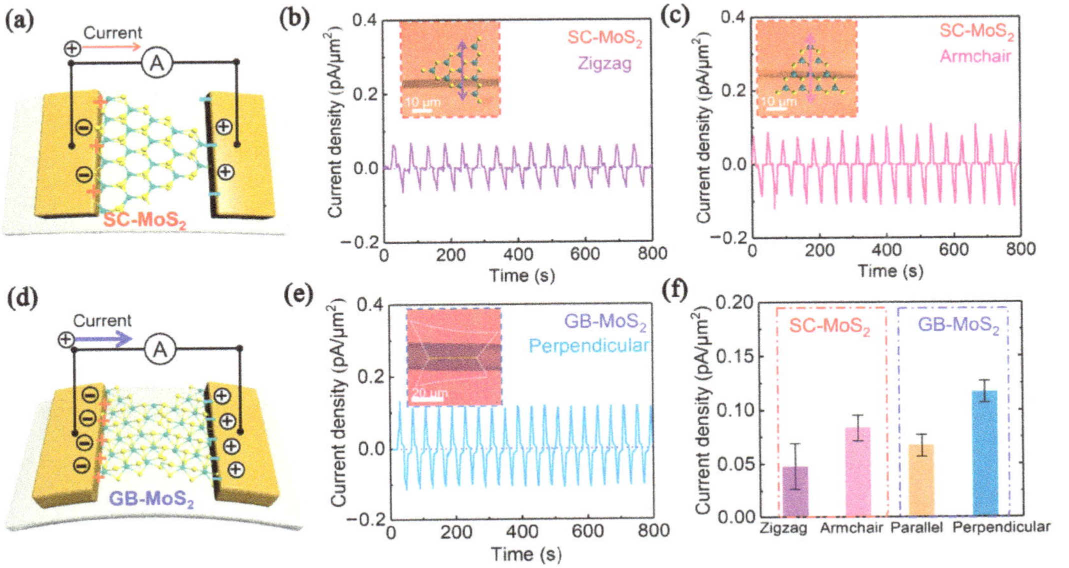

Figure 12. Schematic of a piezoelectric device based on (**a**) single-crystal MoS_2 flake and (**d**) grain boundaries of MoS_2 flake. Recorded current outputs of the SC-MoS_2 devices are (**b**) zigzag and (**c**) armchair, and that of the (**e**) GB-MoS_2 piezoelectric device is perpendicular. (**f**) Statistical data for the current density of different monolayer MoS_2-based piezoelectric devices (reprinted/adapted with permission from Ref. [26]. Copyright 2020 American Chemical Society).

7. Synaptic Transistors

After the concept of artificial intelligence appeared, neuromorphic electronics, which simulate human brain function and information processing, have been proposed as an effective method to solve complex data processing problems [88]. The foundation of this technology is to make artificial synapses, which have low power consumption, small size, and simple structure. A MoS_2 material is appropriate for the construction of artificial synapses due to its excellent electrical properties and optical response. At present, a MoS_2-based artificial synapse has been applied in memory devices, programmable logic circuits, and other fields.

Roy et al. fabricated a MoS_2 vertical synaptic transistor with a titanium and Au electrode. The device exhibited extremely low cycle-to-cycle variability and device-to-device variability and stability in the SET voltage and RESET power distributions. The results showed that there are 26 different conductance states in the device, and each state is maintained for at least 300 s. These devices maintained a consistent on/off ratio during the 1000 DC SET–RESET cycles [89]. Wang et al. proposed a phototransistor based on a MoS_2/graphene heterostructure and an integrated triboelectric nanogenerator to simulate mechanical photon artificial synapses. Synaptic plasticity can be realized by modulating the channel conductivity of the phototransistor by regulating the mechanical displacement of a TENG. The simulation results showed that the image recognition accuracy of the artificial neural network was improved by 92% with the help of mechanical plasticizing [90].

Im et al. fabricated multilevel memory based on van der Waals heterostack (HS) N-$MoSe_2$/N-MoS_2 FETs and extended it to synaptic memory. Figure 13a shows that the synaptic stacked channel FET was used to simulate a biological synapse. Figure 13b,c illustrates that using a voltage pulse could simulate synaptic behavior. HS memory utilized the capture/de-capture phenomenon induced by V_{GS} for programming/erasing functions. Due to the existence of a heterojunction energy barrier between MoS_2 and MoS_2, it could maintain a long retention time of 10^4 s. Based on the P-D characteristics of the device under multiple 60 s short V_{GS} pulses, the simulated recognition rate could reach 94% on average [27], as shown in Figure 13.

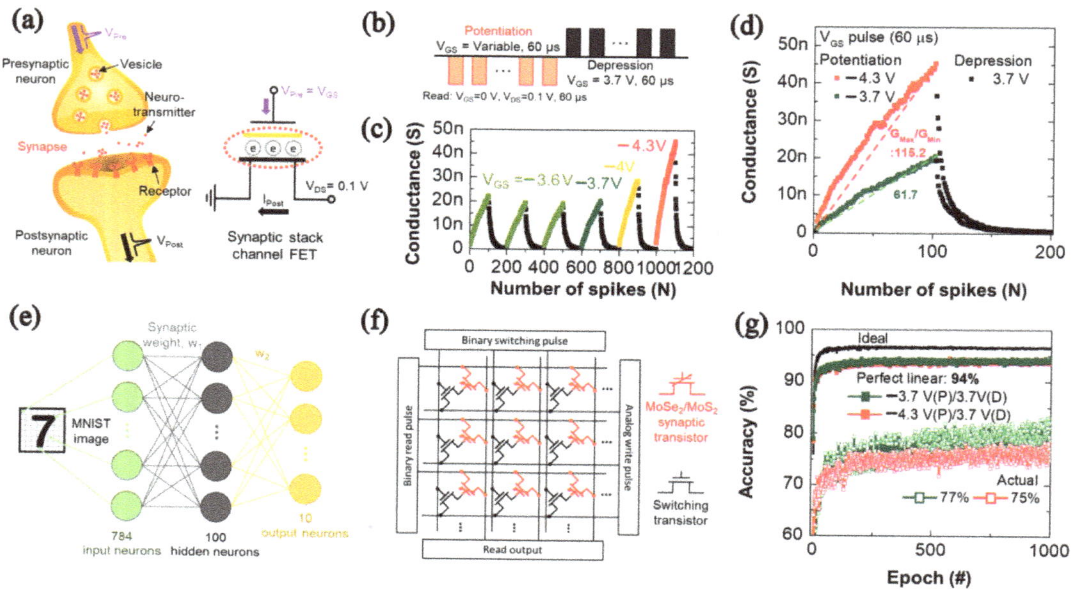

Figure 13. Synaptic memory behavior of stack channel FETs. (**a**) Schematic of a biological synapse in neuron system and synaptic stack channel FETs for neuromorphic function. (**b**) Illustration of a V_{Pre} pulse train. (**c**) Monitored P-D plots vs. the diverse amplitude. (**d**) Two P-D conductance plots with G_{Max}/G_{Min} values. (**e**) Schematics of multilayer perceptron neural network for classification of MNIST handwritten digits. (**f**) Circuit diagram of artificial neural network. (**g**) Simulation results vs. actual P-D behavior. (Reprinted/adapted with permission from Ref. [27]. Copyright 2022 American Chemical Society).

The successful fabrication of artificial synapses based on MoS_2 has proved the possibility of its application in non-von Neumann computing. MoS_2 synaptic transistors with high performance, low power, and large-scale integration characteristics will be widely used in the brain-like chip, logic circuit, and simulated artificial neuromorphic system. However, there are still significant challenges in the fabrication technology and the structure of MoS_2 synaptic transistors.

8. Conclusions

Two-dimensional MoS_2 FETs have attracted wide and in-depth attention as a suitable candidate for optoelectronic devices and next-generation large-size flexible electronics. This benefited from the natural atomic layer thickness and large specific surface area of MoS_2, but it can impact the influence quality, which mainly contains a contact interface and dielectric interface, and they both can then influence the performance of the MoS_2 FET. Thus, in order to obtain higher-performance MoS_2 FETs, it is necessary to focus on its main performance improvement strategies, including optimizing the contact behavior, regulating the conductive channel, and rationalizing the dielectric layer.

To optimize the contact behavior of MoS_2 FETs, this paper reviewed metal, graphene, Mxene, and other new electrodes materials contacting with MoS_2, which can adjust the Fermi level and reduce the Schottky barrier. Self-assembling monolayer functionalized electrodes are also a novel way to improve the contact barrier, which can significantly reduce the I_{on}/I_{off} required by the device. In addition, a 2D/2D van der Waals contact is also the ideal way to optimize the contact behavior. Meanwhile, the intrinsic MoS_2 is a typical N-type semiconductor, so it is difficult to prepare high-performance P-type MoS_2 FETs. At the same time, the defect sites on the surface of the film will produce high resistance and hinder its electrical performance. Therefore, doping is an effective method to improve the performance of conductive channels. Metal doping and non-metal doping can improve the device mobility. More importantly, several different doping methods, such as tantalum, niobium, and He^+ irradiation methods, can realize the transformation of a MoS_2 FET from N-type to P-type. In addition, surface charge accumulation and leakage current due to trap charges are also key challenges for MoS_2 FETs. A reasonable dielectric layer is significant to reduce leakage current and optimize stability. Although the high-K dielectric is a common solution, it can also produce trap charge. Doping dielectric layers, plasma treatment, double-K dielectric layers, etc., are proposed to improve the performance of the dielectric layer.

Functional 2D MoS_2 FETs have been widely used in key and emerging fields such as logic, RF circuits, optoelectronic devices, biosensors, piezoelectric devices, and synaptic transistors. However, there are some important challenges. For example, how to obtain high-performance P-type MoS_2 FETs on a large scale and build complementary circuits in logic circuits. The strong photoluminescence properties of the MoS_2 material promote its application in optoelectronic devices, but its effective mechanism in photoluminescence devices and electroluminescence devices remains to be explored, and wearable high-performance biosensors are of great help to solve the current global epidemic detection. Thus, a MoS_2 artificial synapse can be desired for real-time monitoring. Although its research has been very hot in recent years, there is still a long way to go in the application of a brain-like microarray and artificial neuromorphology. However, we still believe that molybdenum sulfide FETs can be used to construct novel functionalized devices and even super-large scale flexible electronic systems.

Two-dimensional MoS_2 is a very suitable channel layer material for the high-performance FETs, and 2D MoS_2 FETs are the promising candidate for downscaling electronics with short channels, low thickness, small volume, high speed, high sensitivity, light weight, etc. Especially, a MoS_2 FET array based on a high-quality MoS_2 channel will have a large degree of application in the next generation of integrated circuits and flexible electronics. However, monolayer MoS_2 can better meet the high-performance requirements of short channels and even ultra-short-channel FETs. Thus, it is of great practical significance to develop the preparation process of monolayer MoS_2 and MoS_2 FETs performance. Thus, we believe it to be necessary to continue to extensively study in-depth fewer-layer MoS_2 and MoS_2-based FETs.

Author Contributions: C.W. conceived the theme and framework of the review; C.W., Y.S. and H.H. co-authored the manuscript. All authors have read and agreed to the published version of the manuscript.

Funding: This work was supported by the National Natural Science Foundation of China (No. 11604252), the Program of State Key Laboratory of Infrared Physics (No. M202001), the Program of the Key Laboratory of Artificial Microstructure, Ministry of Education (No. 13022019af002), the Science and Technology Program of Shaanxi Province (No. 2017JQ1015), and the Start-Up Funds of Xi'an Polytechnic University (No. BS15026).

Institutional Review Board Statement: Not applicable.

Informed Consent Statement: Not applicable.

Data Availability Statement: Not applicable.

Conflicts of Interest: The authors declare no conflict of interest.

References

1. Liu, Y.; Duan, X.; Shin, H.J.; Park, S.; Huang, Y.; Duan, X. Promises and prospects of two-dimensional transistors. *Nature* **2021**, *591*, 43–53. [CrossRef] [PubMed]
2. Jiang, B.; Yang, Z.; Liu, X.; Liu, Y.; Liao, L. Interface engineering for two-dimensional semiconductor transistors. *Nano Today* **2019**, *25*, 122–134. [CrossRef]
3. Flores-Silva, P.A.; Borja-Hernández, C.; Magaña, C.; Acosta, D.R.; Botello-Méndez, A.R.; Serkovic-Loli, L.N. Graphene field effect transistors using TiO_2 as the dielectric layer. *Phys. E Low-Dimens. Syst. Nanostruct.* **2020**, *124*, 114282. [CrossRef]
4. Feng, X.; Huang, X.; Chen, L.; Tan, W.C.; Wang, L.; Ang, K.-W. High Mobility Anisotropic Black Phosphorus Nanoribbon Field-Effect Transistor. *Adv. Funct. Mater.* **2018**, *28*, 1801524. [CrossRef]
5. Liu, B.; Ma, Y.; Zhang, A.; Chen, L.; Abbas, A.N.; Liu, Y.; Shen, C.; Wan, H.; Zhou, C. High-Performance WSe_2 Field-Effect Transistors via Controlled Formation of In-Plane Heterojunctions. *ACS Nano* **2016**, *10*, 5153–5160. [CrossRef] [PubMed]
6. Kumar, J.; Kuroda, M.A.; Bellus, M.Z.; Han, S.-J.; Chiu, H.-Y. Full-range electrical characteristics of WS_2 transistors. *Appl. Phys. Lett.* **2015**, *106*, 123508. [CrossRef]
7. Liu, H.; Ye, P.D. Dual-Gate MOSFET With Atomic-Layer-Deposited as Top-Gate Dielectric. *IEEE Electron Device Lett.* **2012**, *33*, 546–548. [CrossRef]
8. Larentis, S.; Fallahazad, B.; Tutuc, E. Field-effect transistors and intrinsic mobility in ultra-thin $MoSe_2$ layers. *Appl. Phys. Lett.* **2012**, *101*, 223104. [CrossRef]
9. Liu, Y.; Zhao, S.; Luo, L.; Wang, J.; Zhu, Z.; Xiang, Q.; Deng, Y.; Zhao, Z. Mesenchymal stem cell-derived exosomes ameliorate erection by reducing oxidative stress damage of corpus cavernosum in a rat model of artery injury. *J. Cell Mol. Med.* **2019**, *23*, 7462–7473. [CrossRef]
10. Huang, Y.; Pan, Y.H.; Yang, R.; Bao, L.H.; Meng, L.; Luo, H.L.; Cai, Y.Q.; Liu, G.D.; Zhao, W.J.; Zhou, Z.; et al. Universal mechanical exfoliation of large-area 2D crystals. *Nat. Commun.* **2020**, *11*, 2453. [CrossRef]
11. Xu, X.; Zhang, C.; Hota, M.K.; Liu, Z.; Zhang, X.; Alshareef, H.N. Enhanced Quality of Wafer-Scale MoS_2 Films by a Capping Layer Annealing Process. *Adv. Funct. Mater.* **2020**, *30*, 1908040. [CrossRef]
12. Chang, M.C.; Ho, P.H.; Tseng, M.F.; Lin, F.Y.; Hou, C.H.; Lin, I.K.; Wang, H.; Huang, P.P.; Chiang, C.H.; Yang, Y.C.; et al. Fast growth of large-grain and continuous MoS_2 films through a self-capping vapor-liquid-solid method. *Nat. Commun.* **2020**, *11*, 3682. [CrossRef] [PubMed]
13. Xu, X.; Wang, Z.; Lopatin, S.; Quevedo-Lopez, M.A.; Alshareef, H.N. Wafer scale quasi single crystalline MoS_2 realized by epitaxial phase conversion. *2D Mater.* **2018**, *6*, 015030. [CrossRef]
14. Han, B.; Zhao, Y.; Ma, C.; Wang, C.; Tian, X.; Wang, Y.; Hu, W.; Samori, P. Asymmetric Chemical Functionalization of Top-Contact Electrodes: Tuning the Charge Injection for High-Performance MoS_2 Field-Effect Transistors and Schottky Diodes. *Adv. Mater.* **2022**, *34*, e2109445. [CrossRef] [PubMed]
15. Seo, D.; Lee, D.Y.; Kwon, J.; Lee, J.J.; Taniguchi, T.; Watanabe, K.; Lee, G.-H.; Kim, K.S.; Hone, J.; Kim, Y.D.; et al. High-performance monolayer MoS_2 field-effect transistor with large-scale nitrogen-doped graphene electrodes for Ohmic contact. *Appl. Phys. Lett.* **2019**, *115*, 012104. [CrossRef]
16. Li, H.; Cheng, M.; Wang, P.; Du, R.; Song, L.; He, J.; Shi, J. Reducing Contact Resistance and Boosting Device Performance of Monolayer MoS_2 by In Situ Fe Doping. *Adv. Mater.* **2022**, *34*, e2200885. [CrossRef]
17. Zhang, Z.; Su, M.; Li, G.; Wang, J.; Zhang, X.; Ho, J.C.; Wang, C.; Wan, D.; Liu, X.; Liao, L. Stable Hysteresis-Free MoS_2 Transistors With Low-k/High-k Bilayer Gate Dielectrics. *IEEE Electron Device Lett.* **2020**, *41*, 1036–1039. [CrossRef]
18. Liu, Y.; Gu, F. A wafer-scale synthesis of monolayer MoS_2 and their field-effect transistors toward practical applications. *Nanoscale Adv.* **2021**, *3*, 2117–2138. [CrossRef]
19. Yang, P.; Zou, X.; Zhang, Z.; Hong, M.; Shi, J.; Chen, S.; Shu, J.; Zhao, L.; Jiang, S.; Zhou, X.; et al. Batch production of 6-inch uniform monolayer molybdenum disulfide catalyzed by sodium in glass. *Nat. Commun.* **2018**, *9*, 979. [CrossRef]
20. Liu, F.; Wu, W.; Bai, Y.; Chae, S.H.; Li, Q.; Wang, J.; Hone, J.; Zhu, X.-Y. Disassembling 2D van der Waals crystals into macroscopic monolayers and reassembling into artificial lattices. *Science* **2020**, *367*, 903–906. [CrossRef]
21. Shen, P.C.; Su, C.; Lin, Y.; Chou, A.S.; Cheng, C.C.; Park, J.H.; Chiu, M.H.; Lu, A.Y.; Tang, H.L.; Tavakoli, M.M.; et al. Ultralow contact resistance between semimetal and monolayer semiconductors. *Nature* **2021**, *593*, 211–217. [CrossRef] [PubMed]
22. Kim, J.Y.; Park, H.J.; Lee, S.H.; Seo, C.; Kim, J.; Joo, J. Distinctive Field-Effect Transistors and Ternary Inverters Using Cross-Type WSe_2/MoS_2 Heterojunctions Treated with Polymer Acid. *ACS Appl. Mater. Interfaces* **2020**, *12*, 36530–36539. [CrossRef] [PubMed]
23. Krasnozhon, D.; Lembke, D.; Nyffeler, C.; Leblebici, Y.; Kis, A. MoS_2 transistors operating at gigahertz frequencies. *Nano Lett.* **2014**, *14*, 5905–5911. [CrossRef] [PubMed]
24. Jain, S.K.; Low, M.X.; Taylor, P.D.; Tawfik, S.A.; Spencer, M.J.S.; Kuriakose, S.; Arash, A.; Xu, C.; Sriram, S.; Gupta, G.; et al. 2D/3D Hybrid of MoS_2/GaN for a High-Performance Broadband Photodetector. *ACS Appl. Electron. Mater.* **2021**, *3*, 2407–2414. [CrossRef]
25. Zhang, Y.; Feng, D.; Xu, Y.; Yin, Z.; Dou, W.; Habiba, U.E.; Pan, C.; Zhang, Z.; Mou, H.; Deng, H.; et al. DNA-based functionalization of two-dimensional MoS_2 FET biosensor for ultrasensitive detection of PSA. *Appl. Surf. Sci.* **2021**, *548*, 149169. [CrossRef]
26. Dai, M.; Zheng, W.; Zhang, X.; Wang, S.; Lin, J.; Li, K.; Hu, Y.; Sun, E.; Zhang, J.; Qiu, Y.; et al. Enhanced Piezoelectric Effect Derived from Grain Boundary in MoS_2 Monolayers. *Nano Lett.* **2020**, *20*, 201–207. [CrossRef]

27. Jeong, Y.; Lee, H.J.; Park, J.; Lee, S.; Jin, H.-J.; Park, S.; Cho, H.; Hong, S.; Kim, T.; Kim, K.; et al. Engineering $MoSe_2/MoS_2$ heterojunction traps in 2D transistors for multilevel memory, multiscale display, and synaptic functions. *npj 2D Mater. Appl.* **2022**, *6*, 23. [CrossRef]
28. Ma, X.; Dai, Y.; Yu, L.; Huang, B. Interface Schottky barrier engineering via strain in metal-semiconductor composites. *Nanoscale* **2016**, *8*, 1352–1359. [CrossRef]
29. Zhong, H.; Quhe, R.; Wang, Y.; Ni, Z.; Ye, M.; Song, Z.; Pan, Y.; Yang, J.; Yang, L.; Lei, M.; et al. Interfacial Properties of Monolayer and Bilayer MoS_2 Contacts with Metals: Beyond the Energy Band Calculations. *Sci. Rep.* **2016**, *6*, 21786. [CrossRef]
30. Zhao, P.; Jin, H.; Lv, X.; Huang, B.; Ma, Y.; Dai, Y. Modified MXene: Promising electrode materials for constructing Ohmic contacts with MoS_2 for electronic device applications. *Phys. Chem. Chem. Phys.* **2018**, *20*, 16551–16557. [CrossRef]
31. Lee, I.; Kim, J.N.; Kang, W.T.; Shin, Y.S.; Lee, B.H.; Yu, W.J. Schottky Barrier Variable Graphene/Multilayer-MoS_2 Heterojunction Transistor Used to Overcome Short Channel Effects. *ACS Appl. Mater. Interfaces* **2020**, *12*, 2854–2861. [CrossRef] [PubMed]
32. Zhang, C. Interfacial assembly of two-dimensional MXenes. *J. Energy Chem.* **2021**, *60*, 417–434. [CrossRef]
33. Armin Vahidmohammadi, J.R.Y.G. The world of two-dimensional carbides and nitrides (MXenes). *Science* **2021**, *e372*, abf1581. [CrossRef]
34. Zhou, Y.; Zhuge, X.; An, P.; Du, S. First-principles investigations on MXene-blue phosphorene and MXene-MoS_2 transistors. *Nanotechnology* **2020**, *31*, 395203. [CrossRef] [PubMed]
35. Kim, C.; Lee, K.Y.; Moon, I.; Issarapanacheewin, S.; Yoo, W.J. Metallic contact induced van der Waals gap in a MoS_2 FET. *Nanoscale* **2019**, *11*, 18246–18254. [CrossRef]
36. Manish Chhowalla, D.J.; Zhang, H. Two-dimensional semiconductors for transistors. *Nat. Rev. Mater.* **2016**, *1*, 16052. [CrossRef]
37. Kim, E.; Lee, Y.; Ko, C.; Park, Y.; Yeo, J.; Chen, Y.; Choe, H.S.; Allen, F.I.; Rho, J.; Tongay, S.; et al. Tuning the optical and electrical properties of MoS_2 by selective Ag photo-reduction. *Appl. Phys. Lett.* **2018**, *113*, 013105. [CrossRef]
38. Wei, Z.; Tang, J.; Li, X.; Chi, Z.; Wang, Y.; Wang, Q.; Han, B.; Li, N.; Huang, B.; Li, J.; et al. Wafer-Scale Oxygen-Doped MoS_2 Monolayer. *Small Methods* **2021**, *5*, e2100091. [CrossRef]
39. Li, M.; Wu, X.; Guo, W.; Liu, Y.; Xiao, C.; Ou, T.; Zheng, Y.; Wang, Y. Controllable p-type doping of monolayer MoS_2 with tantalum by one-step chemical vapor deposition. *J. Mater. Chem. C* **2022**, *10*, 7662–7673. [CrossRef]
40. Ma, Z.; Zhang, L.; Zhou, C.; Chan, M. High Current Nb-Doped P-Channel MoS_2 Field-Effect Transistor Using Pt Contact. *IEEE Electron Device Lett.* **2021**, *42*, 343–346. [CrossRef]
41. Han, S.W.; Yun, W.S.; Kim, H.; Kim, Y.; Kim, D.H.; Ahn, C.W.; Ryu, S. Hole doping effect of MoS_2 via electron capture of He(+) ion irradiation. *Sci. Rep.* **2021**, *11*, 23590. [CrossRef]
42. Park, W.; Pak, Y.; Jang, H.Y.; Nam, J.H.; Kim, T.H.; Oh, S.; Choi, S.M.; Kim, Y.; Cho, B. Improvement of the Bias Stress Stability in 2D MoS_2 and WS_2 Transistors with a TiO_2 Interfacial Layer. *Nanomaterials* **2019**, *9*, 1155. [CrossRef] [PubMed]
43. Siao, M.D.; Shen, W.C.; Chen, R.S.; Chang, Z.W.; Shih, M.C.; Chiu, Y.P.; Cheng, C.M. Two-dimensional electronic transport and surface electron accumulation in MoS_2. *Nat. Commun.* **2018**, *9*, 1442. [CrossRef] [PubMed]
44. Jeong, Y.; Jin, H.J.; Park, J.H.; Cho, Y.; Kim, M.; Hong, S.; Jo, W.; Yi, Y.; Im, S. Low Voltage and Ferroelectric 2D Electron Devices Using Lead-Free BaxSr1-xTiO_3 and MoS_2 Channel. *Adv. Funct. Mater.* **2019**, *30*, 1908210. [CrossRef]
45. Kolla, L.G.; Bhattacharjee, S.; S, M.; Bhat, N. High Performance HfO_2 Back Gated Multilayer MoS_2 transistors. *IEEE Electron Device Lett.* **2016**, *37*, 1. [CrossRef]
46. Yu, Z.; Ong, Z.-Y.; Li, S.; Xu, J.-B.; Zhang, G.; Zhang, Y.-W.; Shi, Y.; Wang, X. Analyzing the Carrier Mobility in Transition-Metal Dichalcogenide MoS_2Field-Effect Transistors. *Adv. Funct. Mater.* **2017**, *27*, 1604093. [CrossRef]
47. Ma, N.; Jena, D. Charge Scattering and Mobility in Atomically Thin Semiconductors. *Phys. Rev. X* **2014**, *4*, 011043. [CrossRef]
48. Yu, Z.; Ong, Z.Y.; Pan, Y.; Cui, Y.; Xin, R.; Shi, Y.; Wang, B.; Wu, Y.; Chen, T.; Zhang, Y.W.; et al. Realization of Room-Temperature Phonon-Limited Carrier Transport in Monolayer MoS_2 by Dielectric and Carrier Screening. *Adv. Mater.* **2016**, *28*, 547–552. [CrossRef] [PubMed]
49. Song, X.; Xu, J.; Liu, L.; Lai, P.-T. Comprehensive investigation on CF_4/O_2-plasma treating the interfaces of stacked gate dielectric in MoS_2 transistors. *Appl. Surf. Sci.* **2021**, *542*, 148437. [CrossRef]
50. Song, X.; Xu, J.; Liu, L.; Deng, Y.; Lai, P.T.; Tang, W.M. Optimizing Al-doped ZrO2 as the gate dielectric for MoS_2 field-effect transistors. *Nanotechnology* **2020**, *31*, 135206. [CrossRef]
51. Zhao, X.-Y.; Xu, J.-P.; Liu, L.; Li, Z. Improved Electrical Properties of Top-Gate MoS_2 Transistor With NH3-Plasma Treated HfO_2 as Gate Dielectric. *IEEE Electron Device Lett.* **2020**, *41*, 1364–1367. [CrossRef]
52. Li, W.; Zhou, J.; Cai, S.; Yu, Z.; Zhang, J.; Fang, N.; Li, T.; Wu, Y.; Chen, T.; Xie, X.; et al. Uniform and ultrathin high-κ gate dielectrics for two-dimensional electronic devices. *Nat. Electron.* **2019**, *2*, 563–571. [CrossRef]
53. Zeng, S.; Tang, Z.; Liu, C.; Zhou, P. Electronics based on two-dimensional materials: Status and outlook. *Nano Res.* **2020**, *14*, 1752–1767. [CrossRef]
54. Liu, L.; Lu, Y.; Guo, J. On Monolayer MoS_2 Field-Effect Transistors at the Scaling Limit. *IEEE Trans. Electron Devices* **2013**, *60*, 4133–4139. [CrossRef]
55. Wang, H.; Yu, L.; Lee, Y.H.; Shi, Y.; Hsu, A.; Chin, M.L.; Li, L.J.; Dubey, M.; Kong, J.; Palacios, T. Integrated circuits based on bilayer MoS(2) transistors. *Nano Lett.* **2012**, *12*, 4674–4680. [CrossRef] [PubMed]

56. Wang, L.; Chen, L.; Wong, S.L.; Huang, X.; Liao, W.; Zhu, C.; Lim, Y.F.; Li, D.; Liu, X.; Chi, D.; et al. Electronic Devices and Circuits Based on Wafer-Scale Polycrystalline Monolayer MoS_2 by Chemical Vapor Deposition. *Adv. Electron. Mater.* **2019**, *5*, 1900393. [CrossRef]
57. Yi, J.; Sun, X.; Zhu, C.; Li, S.; Liu, Y.; Zhu, X.; You, W.; Liang, D.; Shuai, Q.; Wu, Y.; et al. Double-Gate MoS_2 Field-Effect Transistors with Full-Range Tunable Threshold Voltage for Multifunctional Logic Circuits. *Adv. Mater.* **2021**, *33*, e2101036. [CrossRef]
58. Liu, C.; Chen, H.; Hou, X.; Zhang, H.; Han, J.; Jiang, Y.G.; Zeng, X.; Zhang, D.W.; Zhou, P. Small footprint transistor architecture for photoswitching logic and in situ memory. *Nat. Nanotechnol.* **2019**, *14*, 662–667. [CrossRef]
59. Lai, R.; Mei, X.B.; Deal, W.R.; Yoshida, W.; Kim, Y.M.; Liu, P.H.; Lee, J.; Uyeda, J.; Radisic, V.; Lange, M.; et al. Sub 50 nm InP HEMT Device with Fmax Greater than 1 THz. In Proceedings of the 2007 IEEE International Electron Devices Meeting, Washington, DC, USA, 10–12 December 2007; pp. 609–611.
60. Zhu, W.; Low, T.; Wang, H.; Ye, P.; Duan, X. Nanoscale electronic devices based on transition metal dichalcogenides. *2D Mater.* **2019**, *6*, 032004. [CrossRef]
61. Cheng, R.; Jiang, S.; Chen, Y.; Liu, Y.; Weiss, N.; Cheng, H.C.; Wu, H.; Huang, Y.; Duan, X. Few-layer molybdenum disulfide transistors and circuits for high-speed flexible electronics. *Nat. Commun.* **2014**, *5*, 5143. [CrossRef]
62. Gao, Q.; Zhang, Z.; Xu, X.; Song, J.; Li, X.; Wu, Y. Scalable high performance radio frequency electronics based on large domain bilayer MoS_2. *Nat. Commun.* **2018**, *9*, 4778. [CrossRef] [PubMed]
63. Dragoman, M.; Aldrigo, M.; Connolly, J.; Povey, I.M.; Iordanescu, S.; Dinescu, A.; Vasilache, D.; Modreanu, M. MoS_2 radio: Detecting radio waves with a two-dimensional transition metal dichalcogenide semiconductor. *Nanotechnology* **2020**, *31*, 06LT01. [CrossRef] [PubMed]
64. Gao, Q.; Zhang, C.; Yang, K.; Pan, X.; Zhang, Z.; Yang, J.; Yi, Z.; Chi, F.; Liu, L. High-Performance CVD Bilayer MoS_2 Radio Frequency Transistors and Gigahertz Mixers for Flexible Nanoelectronics. *Micromachines* **2021**, *12*, 451. [CrossRef]
65. Zhang, X.; Grajal, J.; Vazquez-Roy, J.L.; Radhakrishna, U.; Wang, X.; Chern, W.; Zhou, L.; Lin, Y.; Shen, P.C.; Ji, X.; et al. Two-dimensional MoS_2-enabled flexible rectenna for Wi-Fi-band wireless energy harvesting. *Nature* **2019**, *566*, 368–372. [CrossRef] [PubMed]
66. Ye, M.; Zhang, D.; Yap, Y. Recent Advances in Electronic and Optoelectronic Devices Based on Two-Dimensional Transition Metal Dichalcogenides. *Electronics* **2017**, *6*, 43. [CrossRef]
67. Long, M.; Wang, P.; Fang, H.; Hu, W. Progress, Challenges, and Opportunities for 2D Material Based Photodetectors. *Adv. Funct. Mater.* **2018**, *29*, 1803807. [CrossRef]
68. Cheng, J.; Wang, C.; Zou, X.; Liao, L. Recent Advances in Optoelectronic Devices Based on 2D Materials and Their Heterostructures. *Adv. Opt. Mater.* **2019**, *7*, 1800441. [CrossRef]
69. Yang, P.; Zhang, Z.; Sun, M.; Lin, F.; Cheng, T.; Shi, J.; Xie, C.; Shi, Y.; Jiang, S.; Huan, Y.; et al. Thickness Tunable Wedding-Cake-like MoS_2 Flakes for High-Performance Optoelectronics. *ACS Nano* **2019**, *13*, 3649–3658. [CrossRef]
70. Zou, J.; Huang, Y.; Wang, W.; Li, C.; Wei, S.; Liu, H.; Luo, L.; Du, W.; Shen, K.; Ren, A.; et al. Plasmonic MXene Nanoparticle-Enabled High-Performance Two-Dimensional MoS_2 Photodetectors. *ACS Appl. Mater. Interfaces* **2022**, *14*, 8243–8250. [CrossRef]
71. Tu, L.; Cao, R.; Wang, X.; Chen, Y.; Wu, S.; Wang, F.; Wang, Z.; Shen, H.; Lin, T.; Zhou, P.; et al. Ultrasensitive negative capacitance phototransistors. *Nat. Commun.* **2020**, *11*, 101. [CrossRef]
72. Hu, Y.; Huang, Y.; Tan, C.; Zhang, X.; Lu, Q.; Sindoro, M.; Huang, X.; Huang, W.; Wang, L.; Zhang, H. Two-dimensional transition metal dichalcogenide nanomaterials for biosensing applications. *Mater. Chem. Front.* **2017**, *1*, 24–36. [CrossRef]
73. Fathi-Hafshejani, P.; Azam, N.; Wang, L.; Kuroda, M.A.; Hamilton, M.C.; Hasim, S.; Mahjouri-Samani, M. Two-Dimensional-Material-Based Field-Effect Transistor Biosensor for Detecting COVID-19 Virus (SARS-CoV-2). *ACS Nano* **2021**, *15*, 11461–11469. [CrossRef]
74. Mojsoska, B.; Larsen, S.; Olsen, D.A.; Madsen, J.S.; Brandslund, I.; Alatraktchi, F.A. Rapid SARS-CoV-2 Detection Using Electrochemical Immunosensor. *Sensors* **2021**, *21*, 390. [CrossRef]
75. Barua, S.; Dutta, H.S.; Gogoi, S.; Devi, R.; Khan, R. Nanostructured MoS_2-Based Advanced Biosensors: A Review. *ACS Appl. Nano Mater.* **2017**, *1*, 2–25. [CrossRef]
76. Yan, L.; Shi, H.; Sui, X.; Deng, Z.; Gao, L. MoS_2-DNA and MoS_2 based sensors. *RSC Adv.* **2017**, *7*, 23573–23582. [CrossRef]
77. Dalila, N.R.; Arshad, M.K.M.; Gopinath, S.C.B.; Nuzaihan, M.N.M.; Fathil, M.F.M. Molybdenum disulfide-gold nanoparticle nanocomposite in field-effect transistor back-gate for enhanced C-reactive protein detection. *Mikrochim. Acta* **2020**, *187*, 588. [CrossRef] [PubMed]
78. Zhao, P.; Chen, Y.; Chen, Y.; Hu, S.; Chen, H.; Xiao, W.; Liu, G.; Tang, Y.; Shi, J.; He, Z.; et al. A MoS_2 nanoflower and gold nanoparticle-modified surface plasmon resonance biosensor for a sensitivity-improved immunoassay. *J. Mater. Chem. C* **2020**, *8*, 6861–6868. [CrossRef]
79. Park, H.; Baek, S.; Sen, A.; Jung, B.; Shim, J.; Park, Y.C.; Lee, L.P.; Kim, Y.J.; Kim, S. Ultrasensitive and Selective Field-Effect Transistor-Based Biosensor Created by Rings of MoS_2 Nanopores. *ACS Nano* **2022**, *16*, 1826–1835. [CrossRef]
80. Zhang, Q.; Zuo, S.; Chen, P.; Pan, C. Piezotronics in two-dimensional materials. *InfoMat* **2021**, *3*, 987–1007. [CrossRef]
81. Peng, Y.; Que, M.; Tao, J.; Wang, X.; Lu, J.; Hu, G.; Wan, B.; Xu, Q.; Pan, C. Progress in piezotronic and piezo-phototronic effect of 2D materials. *2D Mater.* **2018**, *5*, 042003. [CrossRef]
82. Hinchet, R.; Khan, U.; Falconi, C.; Kim, S.-W. Piezoelectric properties in two-dimensional materials: Simulations and experiments. *Mater. Today* **2018**, *21*, 611–630. [CrossRef]

83. Wu, T.; Song, Y.; Shi, Z.; Liu, D.; Chen, S.; Xiong, C.; Yang, Q. High-performance nanogenerators based on flexible cellulose nanofibril/MoS_2 nanosheet composite piezoelectric films for energy harvesting. *Nano Energy* **2021**, *80*, 105541. [CrossRef]
84. Wu, W.; Wang, L.; Li, Y.; Zhang, F.; Lin, L.; Niu, S.; Chenet, D.; Zhang, X.; Hao, Y.; Heinz, T.F.; et al. Piezoelectricity of single-atomic-layer MoS_2 for energy conversion and piezotronics. *Nature* **2014**, *514*, 470–474. [CrossRef] [PubMed]
85. Han, S.A.; Kim, T.H.; Kim, S.K.; Lee, K.H.; Park, H.J.; Lee, J.H.; Kim, S.W. Point-Defect-Passivated MoS_2 Nanosheet-Based High Performance Piezoelectric Nanogenerator. *Adv. Mater.* **2018**, *30*, e1800342. [CrossRef]
86. Zhang, D.; Yang, Z.; Li, P.; Pang, M.; Xue, Q. Flexible self-powered high-performance ammonia sensor based on Au-decorated MoSe2 nanoflowers driven by single layer MoS_2-flake piezoelectric nanogenerator. *Nano Energy* **2019**, *65*, 103974. [CrossRef]
87. Dan, T.; Willatzen, M.; Wang, Z.L. Prediction of strong piezoelectricity in 3R-MoS_2 multilayer structures. *Nano Energy* **2019**, *56*, 512–515. [CrossRef]
88. Zhu, Y.; He, Y.; Jiang, S.; Zhu, L.; Chen, C.; Wan, Q. Indium–gallium–zinc–oxide thin-film transistors: Materials, devices, and applications. *J. Semicond.* **2021**, *42*, 031101. [CrossRef]
89. Krishnaparsad, A.D.; Han, S.S.; Shen, Y.Q.; Chung, H.S.; Bae, T.S.; Yoo, C.; Jung, Y.; Lanza, M.; Roy, T. MoS_2 Synapses with Ultra-low Variability and Their Implementation in Boolean Logic. *ACS Nano* **2022**, *16*, 2866–2876. [CrossRef]
90. Yu, J.; Yang, X.; Gao, G.; Xiong, Y.; Wang, Y.; Han, J.; Chen, Y.; Zhang, H.; Sun, Q.; Wang, Z.L. Bioinspired mechano-photonic artificial synapse based on graphene/MoS_2 heterostructure. *Sci. Adv.* **2021**, *7*, eabd9117. [CrossRef]

nanomaterials

Review

Perovskite Quantum Dots for Emerging Displays: Recent Progress and Perspectives

Xinxin Ren [1,†], Xiang Zhang [1,†], Hongxing Xie [1,2], Junhu Cai [1], Chenhui Wang [1], Enguo Chen [1,2,*], Sheng Xu [1,2], Yun Ye [1,2], Jie Sun [1,2], Qun Yan [1,2] and Tailiang Guo [1,2]

1 National & Local United Engineering Laboratory of Flat Panel Display Technology, College of Physics and Information Engineering, Fuzhou University, Fuzhou 350108, China; rxx7321@163.com (X.R.); 111901140@fzu.edu.cn (X.Z.); xiehongxing1992@163.com (H.X.); 15711526574@163.com (J.C.); ahbcxc@163.com (C.W.); xusheng06090@163.com (S.X.); yeyun07@fzu.edu.cn (Y.Y.); jie.sun@fzu.edu.cn (J.S.); qunfyan@gmail.com (Q.Y.); gtl@fzu.edu.cn (T.G.)
2 Fujian Science & Technology Innovation Laboratory for Optoelectronic Information of China, 2 Xueyuan Road, Fuzhou 350108, China
* Correspondence: ceg@fzu.edu.cn; Tel.: +86-13599399819
† These authors contributed equally to this work.

Abstract: The excellent luminescence properties of perovskite quantum dots (PQDs), including wide excitation wavelength range, adjustable emission wavelength, narrow full width at half maximum (FWHM), and high photoluminescence quantum yield (PLQY), highly match the application requirements in emerging displays. Starting from the fundamental structure and the related optical properties, this paper first introduces the existing synthesis approaches of PQDs that have been and will potentially be used for display devices, and then summarizes the stability improving approaches with high retention of PQDs' optical performance. Based on the above, the recent research progress of PQDs in displays is further elaborated. For photoluminescent display applications, the PQDs can be embedded in the backlighting device or color filter for liquid crystal displays (LCD), or they may function as the color conversion layer for blue organic light-emitting diodes (OLED) and blue micro-scale light-emitting diodes (μLED). In terms of next-generation electroluminescent displays, notable progress in perovskite quantum-dot light emitting diodes (PeQLED) has been achieved within the past decade, especially the maximum external quantum efficiency (EQE). To conclude, the key directions for future PQD development are summarized for promising prospects and widespread applications in display fields.

Keywords: perovskite; quantum dots; synthesis; photoluminescent displays; electroluminescent displays

1. Introduction

Photoluminescence (PL) and electroluminescence (EL) are two fundamental excitation modes for current self-emissive displays. EL displays are commonly driven by a given electron current [1,2], whereas the PL type [3] is enabled by down-conversion luminescence materials [4], such as phosphors [5] and quantum dots (QDs) [6,7]. These luminescence materials can be flexibly integrated into light sources, backlight components, functional films, or display panels. They, to a great extent, determine the display performance, and thus need to provide high luminescence efficiency, high color purity, and good stability. The well-known YAG:Ce^{3+} was initially used as the basic material of display phosphor for backlighting devices [8]. However, the wide emission width of this phosphor provides a limited color gamut of merely ~72% according to the National Television System Committee (NTSC) standard [9]. This can be markedly improved by narrow-band red and green-emitting phosphors [10–12], such as the popular green phosphor Beta SiAlON:Eu^{2+}

(~525 nm/~50 nm), red phosphor K2SiF6:Mn^{4+} (~630 nm/~5 nm), etc. Even so, they are still not qualified for next-generation high-resolution and wide-color-gamut displays.

In display fields, a new luminescence material is gradually replacing phosphors due to its narrow emission width, saturated color, and tunable emission [13,14]. This material is named QDs because of its quantum confinement effect and nanoscale dimension [15]. There are roughly three kinds of QD materials for display application: II–VI semiconductor QDs, III–V semiconductor QDs, including perovskite QDs (PQDs), of which the corresponding representatives are shown in Figure 1a–c. Cadmium chalcogenide CdSe-based QDs with a core-shell structure have been successfully commercialized in displays because of their high PL quantum yield (PLQY, ~100%), color quality (FWHM, ~20 nm), and good stability [16,17]. However, the constituent toxic element, Cd, has raised environmental and health concerns. For this reason, InP-based QDs with comparable efficiency and slightly lower color purity have been developed as an alternative [18,19]. To guarantee their performance, the typical InP QD structure is required, which consists of an InP/ZnSe/ZnS core–buffer shell–outer shell structure. This causes the complex and time-consuming synthesis process. In addition, raw materials, especially phosphorus precursor, are expensive.

As a novel kind of QDs, PQDs has gained increasing attention. Compared with the above-mentioned QDs, PQDs exhibit inherent superiorities, including excellent luminescence performance, ease of synthesis, and conveniently tunable emission [20]. These characteristics originate from their unique structural attributes and make them a promising candidate in current and future displays. However, their intrinsic drawbacks also deserve close attention, especially instability and large-scale synthesis [21]. These unresolved issues hinder the breakthroughs in the display industry.

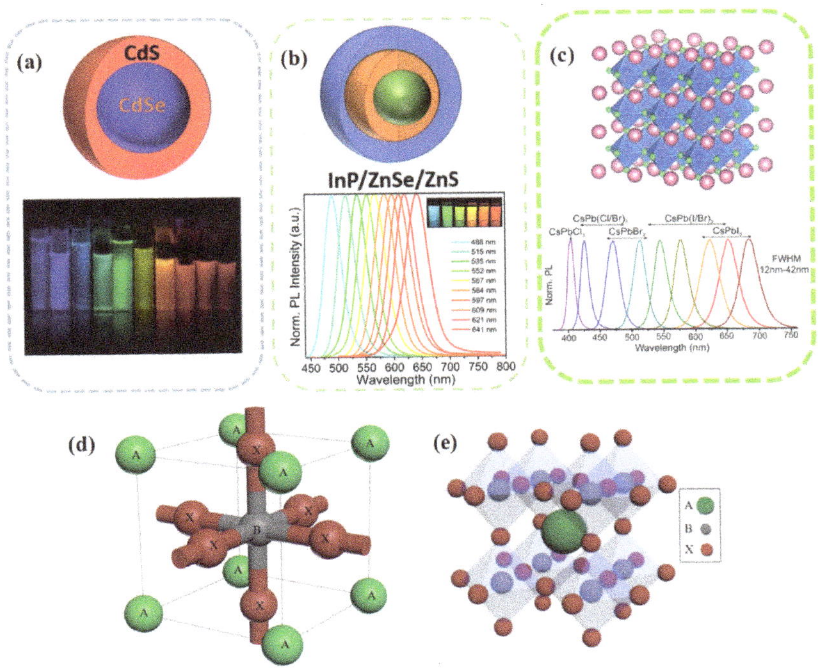

Figure 1. (**a**) CdSe-based QDs. (**b**) InP-based QDs. (**c**) PQDs. Adapted with permission from [22]. Copyright 2020, American Chemical Society. (**d**) The octahedral unit and (**e**) network structure of perovskite ABX_3.

Therefore, this review focuses on the critical demands in display fields, and provides a comprehensive discussion to bridge the gap between fundamental knowledge and display applications. The key content of this review can be divided into three parts. Firstly, the basic synthesis methods of PQDs are concluded, which are deeply discussed from the perspective of PQD's structure and performance. Secondly, we summarize and clarify the strategies for the stability improvement of PQDs in display applications, which is the most difficult issue for display applications. Finally, the latest research progress of PQDs in display fields is described, and the future development direction of PQDs and their display applications are prospected. Beyond other reviews on PQDs, this review provides inspiration for PQDs optimization in display fields, and promotes their application in current and future displays.

2. Fundamental Structure and Optical Properties of PQDs

2.1. Fundamental Structure of PQDs

Halide perovskites have a general formula of ABX_3, where A and B are, respectively, monovalent and divalent cations, and X is a monovalent halide (Cl, Br, I) anion. The basic structural unit of metal halide perovskites is shown in Figure 1d,e, where B-site cations, usually Pb and Sn, will form inorganic octahedra with the six surrounding halide ions. However, when the B site is a mixed cation such as B^{+} and B'^{3+}, the whole structure will form a double-calcite structure with larger crystals. Cubic-phase perovskite (the most regular perovskite) has a corner-sharing structure, which means that the cation at site A is shared by eight neighboring cells, with the location of A at the apex of the cell [23].

The common metal halide perovskites can be further classified into either organic–inorganic (hybrid perovskite quantum dots, HPQDs) or all-inorganic perovskite quantum dots (IPQDs), depending on whether the A cation is an organic molecule such as methylammonium ($CH_3NH_3^+$) and formamidinium (FA^+), or an inorganic cation (commonly Cs^+). The optical and electronic properties of perovskites can be tunable by varying the composition of constituted halide ions and the size of the cations [24,25]. In addition, the dimensionality of perovskites can also be used to tune their optical properties, similar to conventional metal chalcogenide semiconductors [26,27]. Moreover, the reported dimensionality of perovskite can range from the 3D to 0D. Compared with the high-dimensional one, the low-dimensional perovskite nanocrystals (NCs) exhibit very high PLQY partly due to their defect tolerance [25–30], high exciton binding energy [31–33], high optical absorption coefficient [34,35], and tunable carrier diffusion length [36–38].

2.2. Optical Properties of PQDs

After PQDs are excited by external energy, the electrons in the valence band leap into the conduction band, and therefore holes are generated in the valence band. The three types of luminescence are as follows [39]. (1) The electrons return to the valence band and recombine with the holes to emit photons. (2) Electrons are trapped by a defect energy band to emit light. (3) Electrons are trapped by a doped energy level and then emit light.

Four basic optical parameters, including emission stability [40], luminous intensity [40–42], color diversity [43,44], and color purity [45], can be used to characterize the luminescence properties of PQDs. It is remarkable that these basic characteristics also determine their application feasibility in display devices. Among them, emission stability is related to the crystal lattice of PQD materials, and luminous intensity is mainly determined by PLQY. Color diversity can be regulated by changing the PQDs' particle size, composition, and type of ionic elements, while color purity is associated with the FWHM of the emission spectrum.

Compared with organic fluorescent dyes and rare-earth-doped phosphors, PQDs show excellent optical properties in the following four aspects: wide excitation wavelength range [22,40], high PLQY [40–42], adjustable emission wavelength [43,44], and narrow emission FWHM [45].

(1). Wide excitation wavelength range. The excitation spectra of both organic fluorescent dyes and rare-earth-doped phosphors are relatively narrow and may require the use of excitation sources in specific bands to obtain a desired emission spectrum. By contrast, the excitation spectrum of PQDs is continuous and can be excited by arbitrary light higher than the bandgap energy. Therefore, the same excitation light source can simultaneously excite PQDs with different band gaps, resulting in different fluorescent colors.

(2). High PLQY. PQDs with a high molar absorption coefficient have excitation overlap regions so that they can absorb large amounts of excitation light for light conversion. In addition, the defect energy levels caused by internal or surface defects in PQDs can be eliminated by optimizing synthesis methods, modified ligands, and coating, so as to obtain high quantum yield by radiation recombination.

(3). Adjustable emission wavelength. Due to the quantum confinement effect [46], the energy band of the semiconductor is split into discrete energy levels, resulting in different sizes of PQDs having different band gaps. In other words, by regulating the particle size, the light-emitting color of PQDs can be easily tuned to the required wavelength range for various applications. Different from other QDs, the luminescence color of PQDs can also be changed by controlling the components of the halogen anions, which has the potential to cover the entire visible spectrum.

(4). Narrow emission FWHM. For PQDs, the relaxation rate of electrons and holes in the band is much higher than the composite rate of thermal excitons. It is hard to have recombination between high-level electrons and holes. Thus, the luminescence spectra of monodispersed PQDs are basically symmetrical. The FWHM is comparable to the low-energy edge of the first exciton absorption peak, and the luminescence peak energy is slightly lower than the first exciton absorption peak.

Based on the above excellent optical properties, PQDs show outstanding performance to better meet the need in emerging display, as shown in Figure 2.

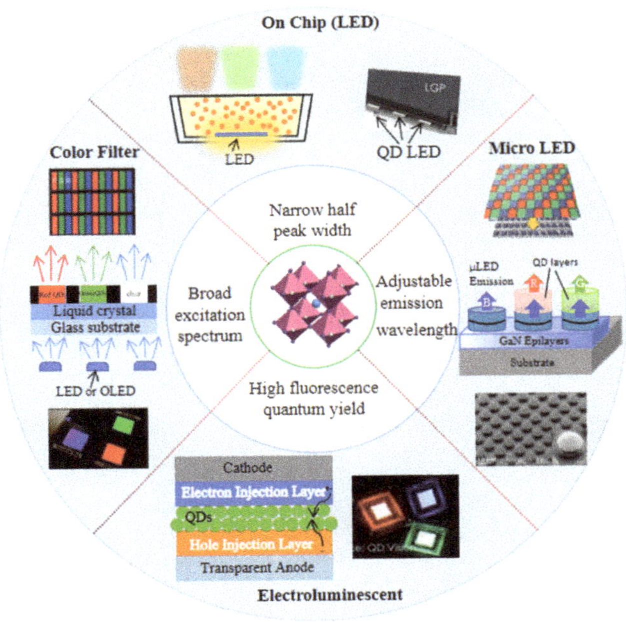

Figure 2. Schematic diagram of the research direction of PQD displays.

3. Synthesis Methods of PQDs

The synthesis of $CsPbX_3$ PQDs showing bright emission and a wide color gamut was first reported by Loredana et al. in 2015, and it is widely known as the hot injection method [47]. The Cs-oleate precursor was prepared in advance, and then injected into a PbX_2 (X = Cl, Br, I) solution dissolved in oleic acid (OA), oleylamine (OAm), and octadecene (ODE) at high temperature and in a nitrogen atmosphere. After a few seconds, the temperature of the reaction system was quickly dropped to room temperature and the PQDs could be obtained via centrifugation. By using this method, the cubic $CsPbX_3$ QDs with a PLQY of 50~90% and an FWHM of 12~42 nm were successfully synthesized (Figure 3a), which paved a new way for the development of perovskite. The hot injection method introduces OA, and OAm ligands, providing potential for subsequent studies of ligand modification. In addition, this method facilitates the introduction of ions into perovskite lattice, laying the foundation for the study of ion doping. In the same year, Nedelcu et al. [48] prepared IPQDs by introducing different halogen elements into $CsPbX_3$ QDs for an anion exchange reaction (Figure 3b), and finally realized the full spectral luminescence (410~700 nm) with the PLQY of 20~80%.

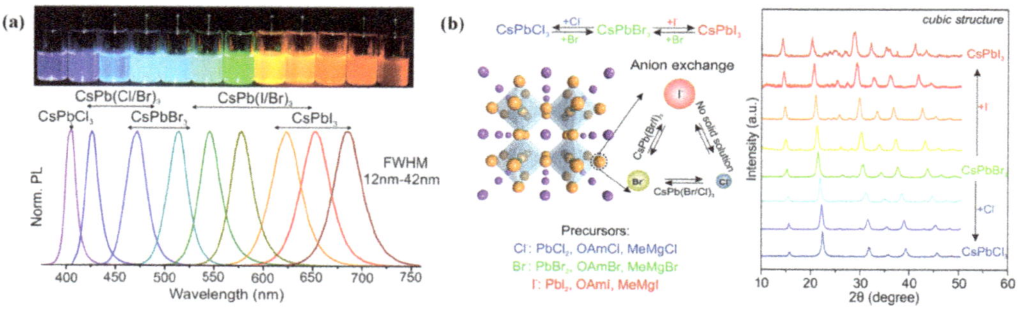

Figure 3. (**a**) Colloidal perovskite $CsPbX_3$ NCs (X = Cl, Br, I) covering the whole visible spectral region with narrow and bright emission. Adapted with Open Access from [26]. Copyright 2015, American Chemical Society. (**b**) The anion exchange diagram and X-ray diffraction diagram of PQDs. Adapted with permission from [48]. Copyright 2015, American Chemical Society.

Due to strict reaction conditions and a complex process, the hot injection method is still difficult for mass production at present. In 2016, the room-temperature reprecipitation method was proposed based on the differences in the solubility of ions in different solvents [49]. OAm and OA as surface ligands and PbX_2 and CsX as ion sources were dissolved in dimethylformamide (DMF) at room temperature to serve as precursors. The precursor at a supersaturated state was injected into the toluene solution and a large number of perovskite crystals were precipitated (Figure 4a). The resulting perovskite had excellent optical properties, with a PLQY of 80%, 95%, and 70% and a FWHM of 35 nm, 20 nm, and 18 nm for red, green, and blue, respectively. In this method, ligands function to passivate the QDs' surface to reduce surface defects and inhibit nonradiative recombination to improve the luminescence performance and lifetime. In addition, it is simple to operate without high temperature and an inert gas environment, and it is less affected by the environment. Therefore, it does have high repeatability compared with the hot injection method. In the same year, Tong et al. [50] described a universal nonpolar solvent ultrasound method which mixed precursors of cesium and lead halide with the end-sealing ligand (OAm and OA) in ODE followed by sonication for 10 min (Figure 4b). The PLQY of the prepared red, green, and blue perovskites, respectively, reached 90%, 92%, and 10%, and the synthesized $CsPbBr_3$ was highly monodisperse. This simple, fast, and ligand-modifiable method is expected to achieve commercial production of perovskite.

In 2017, the microwave-assisted synthesis of $CsPbX_3$ NCs with different morphologies was first reported by Pan et al. [51]. Cesium acetate, lead halide (PbX_2, X = Cl, Br, I or their mixtures), a certain amount of trioctylphosphine oxide (TOPO), OA, OAm, and ODE were mixed in a microwave quartz tube and then put into a microwave reactor. Nanoplates and nanocubes were obtained at low and high reaction temperature, respectively, while nanorods could be formed by pre-dissolving precursors. This method provided uniform particle size distribution, simple operation, no inert gas, and less environmental impact. Compared with the hot injection method, it has a high repeatability and no other pre-treatment. In the same year, the solvothermal method was proposed for the synthesis of IPQDs [52]. Cs_2CO_3 and PbX_2, used as precursors, were mixed with OA, OAm, and ODE in the autoclave and maintained at 160 °C for a while (Figure 4c). $CsPbX_3$ QDs and ultrathin nanowires with uniform cubic phase were successfully prepared with the PLQY reaching 80%. Although this simple preparation method could obtain high-quality IPQDs with controllable morphology, the uncontrollable system temperature made it rarely used in doping strategies and ligand modification.

The mechanochemical synthesis method was first proposed in 2017 [53]. Solid $PbBr_2$, ABr, and capping ligands were mixed and ground at room temperature for a while. Square and rectangular ($CsPbBr_3$), spherical ($MAPbBr_3$), and parallelogram ($FAPbBr_3$) nanoparticles (NPs) were prepared through this solid-phase method. Although its PLQY of 13% was significantly lower than in traditional liquid-phase method, it still showed certain good characteristics such as high yield, simplicity, and fast synthesis process. Due to the solid reaction system, the ligand modification strategy was hardly applied. For this reason, the wet ball milling method for preparing colloidal nanocrystals was further proposed by Kovalenko et al. in 2018 [54], which was composed of $APbBr_3$ mixed with solvent and oil-based ammonium bromide ligand. Figure 4d illustrates the working principle of the wet ball milling method. In 2019, Palazon et al. [55] revealed the process mechanism, dynamics, and possible side effects of dry ball milling. The changes of mechanochemical synthesis with different time variations were studied in detail, and it was found that the drying and rapid (5 min) process affected the excellent phase purity of IPQDs.

In 2018, Guo et al. [56] synthesized $CsPbX_3$ microcrystals using chemical vapor deposition (CVD) at room temperature. The working process was summarized as follows. PbX_2 and CsX (X = Cl, Br, I) were mixed in a reaction chamber. The substrate could be made of sapphire, SiO_2, or Si. The products $CsPbI_3$, $CsPbBr_3$, and $CsPbCl_3$ were grown at 580 °C, 620 °C, and 620 °C, respectively, with argon at the rate of 100 mL/min and the growth time of 30 min. The fluorescence lifetime was 59.7 ns ($CsPbI_3$), 36.9 ns ($CsPbBr_3$), and 3.5 ns ($CsPbCl_3$), respectively. By using this method, the white-light-emitting chips could be successfully prepared on substrates. Though the large size (μm) and high precision of the experimental equipment still limit its application, it does show a certain potential in display backlights due to excellent performance of the crystal product.

In recent years, the synthesis of various kinds of nanocrystals with good homogeneity using a fully automated microfluidic platform has become a hot research topic [57,58]. This microfluidic platform allows the parameters of the synthesized nanocrystals to be varied by changing the precursor molar ratio (e.g., Cs/Pb, FA/Pb, Cs/FA, and Br/I), growth time (determined by flow rate and tube length), and temperature in a systematic and independent way. Droplets are generated by adjusting the flow rates of the carrier phase (50~200 μL/min) and that of the dispersed phase (1.2~50 μL/min) (Figure 4e left), and each droplet can be viewed as a small hot-injection reaction system (Figure 4e right). In 2018, Lignos et al. [58] further explored the synthesis of colloidal QDs in the near infrared using a microfluidic platform. The synthesis results showed that untreated colloidal QDs had an emission FWHM within 45~65 nm. The NCs could further narrow the PL FWHM to 40 nm after a series of post-treatments (e.g., isolation, size selection, and purification), while the synchrotron X-ray scattering clearly showed a cubic structure of $Cs_xFA_{1-x}Pb(Br_{1-y}I_y)_3$ NCs. Finally, the electroluminescent devices prepared by this colloidal QDs obtained a 5.9% EQE.

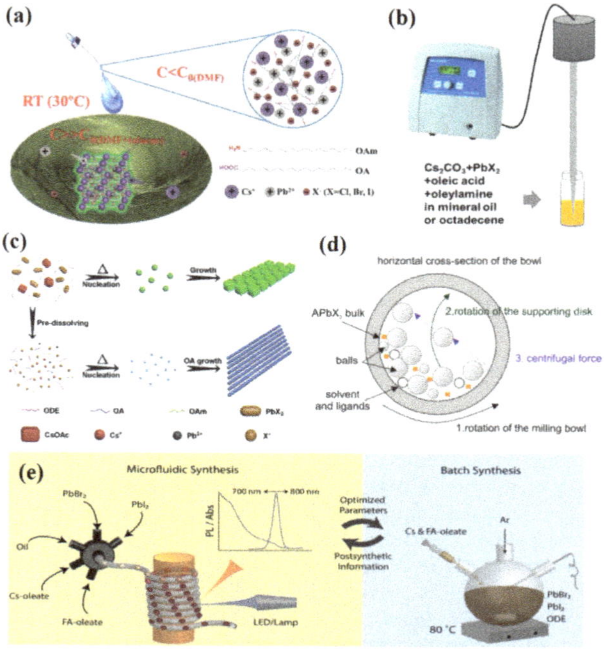

Figure 4. (**a**) Room-temperature precipitation method. Reproduced with permission from [49]. Copyright 2016, Wiley. (**b**) Ultrasonic method. Reproduced with permission from [50]. Copyright 2016, Wiley. (**c**) Solvothermal method. Reproduced with permission from [52]. Copyright 2017, Wiley. (**d**) Wet ball milling method. Adapted with Open Access from [54]. Copyright 2018, American Chemical Society. (**e**) Automatic microfluidic method. Adapted with Open Access from [58]. Copyright 2018, American Chemical Society.

To sum up, efficient, convenient, and low-cost synthesis methods have been proposed for PQDs, which lay a foundation for its potential applications in displays. Table 1 summarizes the characteristics, advantages, and disadvantages of the existing perovskite synthesis methods, with references attached.

Table 1. The existing synthesis methods of PQDs.

Methods	Principle	Results	Drawbacks	Reference
Hot injection	High temperature	High yield, good properties, suitable for ion doping and ligand modification, widely used	Complex process	[26]
Anion exchange	Doping	Full-spectrum luminescence, easy X-position doping	Multi-step process	[47]
Room-temperature re-precipitation	Different solubility	Easy operation, high repeatability, suitable for ligand modification	Uneven size	[48]
Ultrasonic method	Ultrasonic treatment	Easy operation, suitable for ligand modification	High cost	[49]
Microwave-assisted synthesis	Microwave treatment	Easy operation, high repeatability, suitable for ligand modification	High cost	[50]

Table 1. *Cont.*

Methods	Principle	Results	Drawbacks	Reference
Solvothermal synthesis	Mixed high temperature	Easy to synthesize, controllable morphology	System temperature unevenness, not suitable for ion doping and ligand modification	[51]
Mechanochemical synthesis	Mixed grinding	High yield, easy to synthesize	Not applicable to ligand modification	[52]
Wet ball milling	Mixed grinding	Easy to synthesize	Low synthetic efficiency	[53]
Dry ball milling	Mixed grinding, solvent-free	Fast, high synthetic purity	Easy to generate surface defects	[54]
Chemical vapor deposition	Chemical reaction, deposition	Excellent performance	Large size, precise equipment	[55]
Microfluidic platform	Carrier spacing reaction	Automatic, homogeneity	Immature	[57]

4. Performance Improvement of PQDs

Although PQDs show high PLQY, low Auger recombination loss, and large exciton binding energy, the poor humidity resistance and thermal stability of PQDs greatly limit their practical application in the display field. Effective stabilization measures inevitably result in a decline in optical performance. This section focuses on the strategies to improve the stability and optical performance of PQDs.

4.1. Ion Doping of PQDs

An ion doping strategy has become an important way to improve the oxygen/moisture resistance and luminescence properties of PQDs, which initially came from the doping strategy of semiconductor QDs. ABX_3 has three different lattice positions, and different lattice positions have different effects on the material. Taking $CsPbBr_3$ as an example, Cs has little effect on its electronic structure. However, the 4p orbital of Br and the 6p orbital of Pb contribute greatly to the valence band and conduction band of the crystal, respectively. In addition, the excitation and recombination of electrons and holes are carried out in an octahedron [58]. Therefore, inorganic octahedrals are very important for the luminescence of PQDs, and different ionic doping will have different effects on the properties of PQDs.

The A-site has a great influence on the structure and stability of PQDs. Considering the valence distribution of perovskite lattice, monovalent cations (such as BA^+ [59]) are mostly used as A-site doped ions. Because of the strong oxidation resistance, alkali metals (Na^+ [60], K^+ [61,62], Rb^+ [63], etc.) are regarded as ideal doping ions at the A-site. In 2018, Liu et al. [61] doped $CsPbCl_3$ with K^+. The introduction of K^+ reduced the defects of perovskite and narrowed the FWHM of the emission peak. With the increase of doping concentration, the PLQY increased from 3.2% to 7.2%, and the average lifetime was also improved. In the same year, Huang et al. [62] adopted Rb^+ doping and found that when the proportion of Rb^+ was close to 75%, the PLQY of PQDs could be effectively improved. Among them, the PLQY of blue PQDs changed most significantly, from 45% to 86%. Rb^+ doping also effectively improved the UV light stability and thermal stability of perovskite.

The B-site contributes greatly to the conduction band of QDs, which mainly affects their photoelectric properties. The strategy of B-site doping can reduce the lead content to a certain extent, which is very important for PQDs. Commonly used doped ions include Eu^{3+} [61], Bi^{3+} [64,65], Tm^{3+} [66], Cu^{2+} [67], Zn^{2+} [68], Fe^{2+} [69,70], Mn^{2+} [64,66,71] (Figure 5a), CO^{2+} [72], etc. In 2018, Liu et al. [61] doped with Eu^{3+} to increase PLQY from 10.3% to 31.2%. In 2019, Bi et al. [67] doped Cu^{2+} ions into PQDs to improve its thermal stability and optical properties. The prepared PQDs showed bright blue photoluminescence (PL) at 450~460 nm, with a quantum yield of more than 80% and excellent thermal stability (Figure 5b). In 2019, Hu et al. [69] doped with Fe^{2+} to improve the size uniformity and

PLQY of $CsPbCl_3$ QDs, which also could reduce the defect recombination and non-radiation recombination of PQDs and prolong the average fluorescence lifetime. In 2021, Gualdrón-Reyes et al. [73] used 7% Sr instead of Pb to achieve $FAPb_{1-x}Sr_xI_3$ PQDs with 100% PLQY and high stability for 8 months under a relative humidity of 40~50%, and T_{80} = 6.5 months. These are among the highest values reported for halide PQDs under air ambient conditions until now. $FAPb_{0.93}Sr_{0.07}I_3$ PQDs also showed light brightening enhancement under UV irradiation for 12 h and the PLQY recovered to 100% in 15 days after synthesis. Among many doped ions, Mn^{2+} was the most concerned. The doping of Mn^{2+} could bring new energy levels and match the energy band structure of $CsPbX_3$, which can realize the band edge luminescence and doped luminescence of $CsPbX_3$ QDs at the same time [71]. The doped luminescence fluorescence lifetime of Mn^{2+} was up to milliseconds, which was an important basis for display applications.

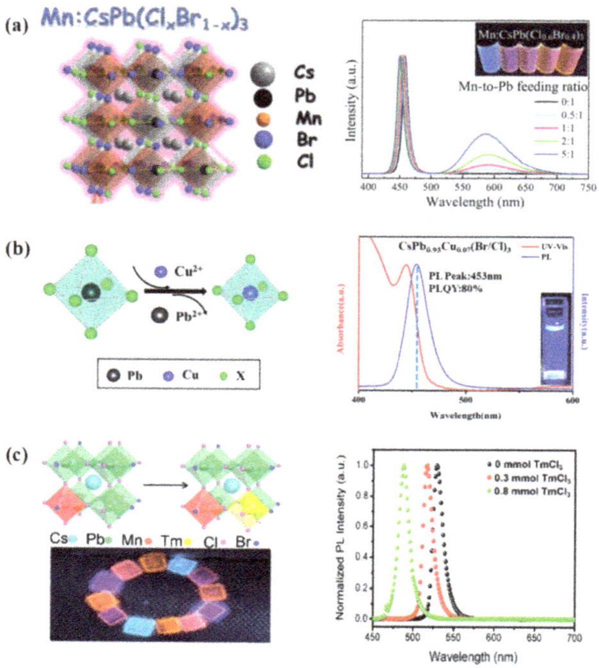

Figure 5. (**a**) Mn^{2+} doping. Reproduced with permission from [71]. Copyright 2019, Wiley. (**b**) Cu^{2+} doping. Adapted with permission from [67]. Copyright 2019, American Chemical Society. (**c**) Tm^{3+} doping. Adapted with permission from [66]. Copyright 2019, American Chemical Society.

Compared with IPQDs, organic perovskite prepared through a solution process could well be integrated into the silicon-based wafer. Organic molecules in perovskite materials show high polarizability, but larger molecules are required in the process of high polarization, and need to increase the hole space of perovskite. Given this, $(DCl)(NH_4)(BF_4)_3$ was synthesized by introducing a BF_4^- group at the X-site, showing a linear photoelectric coefficient of 20 pm V^{-1}, which was 10 times higher than that of metal halide perovskite [74]. The photoelectric response of organic perovskite materials was close to $LiNbO_3$ (Reff $\approx$ 30 pm V^{-1}).

Multiple ion doping is another effective way to adjust exciton dynamics and realize white light emission. $CsPbCl_3$ QDs doped with Bi^{3+} and Mn^{2+} were synthesized by Shao et al. [64]. By strictly controlling the concentration of doped ions, the white light emission was composed of blue QD band edge luminescence, green Bi^{3+} doped luminescence, and red Mn^{2+} doped luminescence. It was the first time to achieve the white light emission with a single component; at the same time, the correlated color temperature (CCT) could be adjusted from 4250 K to 19,000 K. In 2019, Luo et al. [66] synthesized doped $CsPbBr_{2.2}Cl_{0.8}$ QDs with Tm^{3+} and Mn^{2+}, and introduced the $^1G_4$ level of Tm between the conduction band of the QD and the $^4T_1$ level of Mn, which effectively promoted the energy transfer of exciton from the QD body to the Mn^{2+} doped level (Figure 5c). Single component white-light QDs were obtained with a PLQY of 54%. The problem that excessive Mn^{2+} led to a sharp decline in PLQY was solved. Doping Tm^{3+} significantly improved the air stability and thermal stability of QDs, which was instructive for obtaining single-component white-light QDs with high PLQY and further display application.

Table 2 summarizes the luminescence properties of PQDs realized by different types of ion doping, including PL, FWHM, PLQY, and the lifetime after doping.

Table 2. Luminescence properties of different ion-doped PQDs.

Doping	Excitation (nm)	PL (nm)	FWHM (nm)	PLQY (%)	τ (ns)	Stability	Advantages	Reference
						A—site doping		
BA^+	–	–	–	49.44	24.58	Stable (50 days, 80% RH)	Reduced dimensionality	[59]
K^+	365	408	12.7	10.3	13.6	–	Greatly improved PLQY	[61]
Rb^+	365	505–515	18–20	93	5.32	30% (100 °C, 24 h)	Increased exciton binding energy	[63]
						B—site doping		
Eu^{3+}	365	408	11.3	31.2	15.24	–	Greatly improved PLQY	[61]
Bi^{3+}	365	420–520	–	52	9.5	70% (30 days, air)	Lead-free PQDs	[64]
Tm^{3+}	365	–	–	54	4.8–5	Stable (80 °C, 24 h)	Introduction of new energy level	[66]
Cu^{2+}	365	450–460	15–26	>80%	2.3–5	90% (30 days, 60% RH, 25 °C)	Eliminating halide vacancies	[67]
Zn^{2+}	365/380	395–550	47	79.05	–	63.77% PLQY (50 days, air)	Lead-free PQDs	[68]
Fe^{2+}	–	401–403	13.8–14.6	6.2	14.6	–	Size homogeneity improvement	[69]
Mn^{2+}	365	–	–	65	–	–	Toxic ions reduction and PLQY improvement	[71]
Co^{2+}	365	516	18–20	89	17.93	90% (50 days)	Defect passivation	[72]
Sr	–	589, 583, 530	–	100	–	Stable for 8 months (40–50% RH, 6.5 months)	Defect passivation	[73]
						X—site doping		
BF_4^-	–	515	–	–	–	–	Increased hole space of perovskite	[74]
						Multiple ion doping		
Bi^{3+}, Mn^{2+}	365	420–520	–	52	9.5	70% (30 days, air)	Wide range of CCT	[64]
Tm^{3+}, Mn^{2+}	365	–	–	54	4.8–5	Stable (80 °C, 24 h)	Promotion of exciton energy transfer	[66]

4.2. Ligand Modification of PQDs

As mentioned above, OA and OAm are mostly used as ligand materials during the synthesis of PQDs. These long-chain organic ligands are attached to the surface of PQDs and have a strong protective effect on PQDs. However, OA and OAm can easily fall off after proton exchange, resulting in the degradation of the optical properties and color stability of PQDs [75]. In addition, the insulation of long-chain organic ligands will hinder charge transmission, leading to the poor conductivity of the material, which limits its application in the display field. Therefore, choosing appropriate ligands to replace long-chain organic ligands is an important way to ensure the stability and promote the application of PQDs.

The volume and chain length of ligands have a significant impact on the optical properties and stability of PQDs. In 2018, Song et al. [76] utilized the synergy effect with three short ligands: tetraoctylammonium bromide (TOAB), DDAB, and octanoic acid. This effectively enhanced charge injection and transportation in QD films. The highest EQE of the prepared LED devices reached 11.6%. In 2019, SCN^- was used as ligands to modify the surface of $CsPbX_3$ [77], which reduced Pb^{2+} surface defects and improved PLQY. The performance of LED devices prepared by modified IPQDs was improved by 25%. In the same year, Park et al. [78] found that small ligands surrounding the surface of IPQDs could effectively passivate the surface and reduce aggregation (Figure 6a). Finally, the IPQDs optimized by DDAB ligands was used to construct an efficient green LED with the current efficiency (CE) of 31.7 cd/A and the EQE of 9.7%, which is 16 times higher than the perovskite LED with traditional OAm ligands. In 2020, Huang et al. [79] synthesized DDAB-capped $CsPbI_3$ QDs. The introduced DDAB could firmly be combined with the surface of PQDs and effectively passivate surface defects. The DDAB-$CsPbI_3$ NCs retained PLQY > 80% for at least 60 days. In 2021, Li et al. [80] explored the influence of DDAB on the optical performance of $CsPbBr_3$ QDs. They also studied DDAB-$CsPbBr_3$ QDs synthesized from different molar ratios of Pb:DDAB (Figure 6b). In the same year, Yang et al. [81] fabricated $CsPbBr_3$ QDs using a one-step microwave method, where α-phase poly(vinylidene fluoride) (PVDF) acted as the surface-capping ligands. The PLQY of PVDF-$CsPbBr_3$ QDs reached up to 98% (much higher than pristine PQDs with OA/OAm as capping ligands). Moreover, these PQDs had excellent stability in the desired cubic phase structure and enhanced PL stability under ambient conditions.

Replacing organic ligands with inorganic materials is also an effective measure to improve the conductivity of IPQDs. In 2018, Song et al. [82] adopted an organic–inorganic hybrid ligand (OIHL) to passivate IPQDs to control its surface state and subsequently construct an efficient LED device. The inorganic $ZnBr_2$ ligand could enhance the radiation recombination and carrier transport of IPQDs. In addition, it was also found that other metal bromides ($MnBr_2$, $GaBr_3$ and $InBr_3$) could play the same role (Figure 6c). In 2020, Yang et al. [83] introduced K^+ to partially replace organic ligands, which not only inhibited non-radiative recombination to obtain blue $CsPb(Br/Cl)_3$ with the PLQY of 38.4%, but also improved the charge carrier transport performance of IPQDs. They finally obtained a stable and efficient blue perovskite LED with an EQE of 1.96%.

Compared with the direct substitution of ligands, the change in chemical bonds between ligands is also worth considering. Jang et al. [84] found that the formation of chemical bonds could effectively improve the stability of PQDs. They reported extremely stable crosslinked perovskite NPs, in which the unsaturated hydrocarbons in both the acid and base ligands of NPs were chemically crosslinked with a methacrylate-functionalized matrix, preventing decomposition of the perovskite crystals. Counterintuitively, water vapor permeating through the crosslinked matrix could chemically passivate surface defects in the NPs and reduce non-radiative recombination. Green-emitting and white-emitting flexible large-area displays were demonstrated which were stable in air and water for >400 days. The design strategies provided a meaningful breakthrough toward the commercialization of perovskite NPs in display applications.

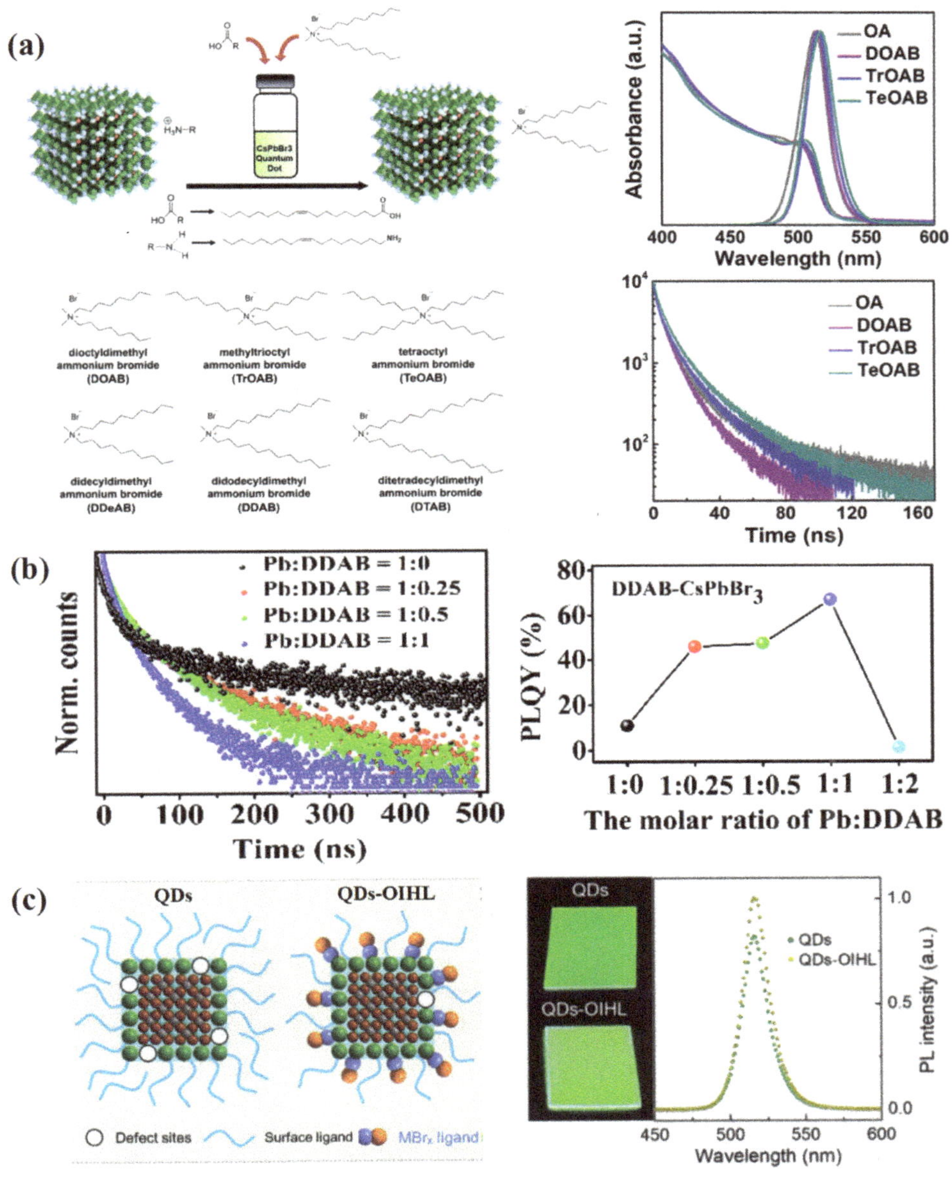

Figure 6. (**a**) Quaternary ammonium bromide ligand materials with different bulkiness and length. Adapted with permission from [78]. Copyright 2019, American Chemical Society. (**b**) PLQY and PL decay curves of DDAB-$CsPbBr_3$ QDs with different molar ratios of Pb: DDAB. Reproduced with permission from [80]. Copyright 2021, ELSEVIER. (**c**) OIHL passivation strategy for IPQDs. Reproduced with permission from [82]. Copyright 2018, Wiley.

4.3. Coating of PQDs

The dynamic characteristics and low lattice energy of PQDs lead to their dissolution in almost all polar solvents, even in water. This problem continues to plague researchers until there is a strategy to completely encapsulate PQDs with inert shell materials. Compared with the ion doping and ligand modification strategies mentioned above, the coating strategy is more straightforward to improve performance by isolating as much water and oxygen as possible. So far, the silica (SiO_2) coating method has been widely used in traditional QDs [85–87], such as lanthanide-doped QDs and magnetic nanocomposites. It is worth noting that SiO_2 is an inorganic oxide with chemical stability and optical transparency within the whole visible spectrum. Coating PQDs with SiO_2 not only retains the optical properties of luminescent materials, but also protects the materials from the dissolution by polar solvents. The silica coating for $CsPbX_3$ QDs is a typical process using Stöber [88] or reverse microemulsion methods [89]. Hu et al. [90] proposed a simple and easy method to grow silica shells on $CsPbX_3$ QDs in situ without using any water. During the preparation process, the amorphous SiO_2 layer was rapidly formed on PQDs through high-temperature injection of the silica precursor tetraethyl orthosilicate (TEOS). Compared with pristine $CsPbBr_3$ QDs, the performance of the prepared $CsPbBr_3/SiO_2$ composites has been significantly improved, such as luminous intensity, nonblinking properties, and optical stability.

In addition to the silica coating strategy, Chen et al. [91] prepared $CsPbX_3/ZnS$ heterostructures using a simple liquid-phase process in which QDs can be maintained in the air for 12 days. Li et al. [92] proposed a method to obtain monodisperse $CsPbBr_3/TiO_2$ core/shell NCs. The prepared monodisperse $CsPbBr_3/TiO_2$ nanocomposites have excellent water stability and PL intensity. In addition, atomic layer deposition (ALD) is an advanced method of depositing single-atom thickness films to achieve high-quality thickness control [93,94]. Xiang et al. [95] coated nanoscale alumina on the $CsPbBr_3$ QDs-silica luminescent sphere via atomic layer deposition (ALD) technology to selectively passivate the surface defect sites of $CsPbBr_3$ QDs (Figure 7a). The inorganic alumina coating layers can effectively reduce the ion migration and crystal deformation of $CsPbBr_3$ QDs.

At present, PQDs can be coated with various mesoporous materials, such as TiO_2, AlO_X, SiO_2, and so on. Wang et al. [96] mixed green $CsPbBr_3$ PQDs with mesoporous silica whose pore size was approximately 12~15 nm, as shown in Figure 7b. In addition, mixing green QD-containing mesoporous silica nanocomposites with red PQDs can prevent the anion-exchange effect, improve thermal and optical stability, and finally realize on-chip LED devices with 113% NTSC. Dirin et al. [97] reported that PQDs were formed in situ assisted with mesoporous silica. High-quality PQDs were obtained by infiltrating the perovskite precursor solution into the mesoporous material and drying. Malgras et al. [98] controlled the growth of the material and found that its emission spectrum under normal environmental and light conditions remained almost unchanged.

Inorganic salts could be another choice to coat PQDs to improve their stability. Yang et al. [99] selected $NaNO_3$ as a coating material for $CH_3NH_3PbBr_3$ QDs. The precursor was dissolved in DMF solvent and then transferred to the toluene solvent. After that, $CH_3NH_3PbBr_3/NaNO_3$ NCs were obtained through a one-step reprecipitation procedure, and their thermal stability and photostability were greatly improved. Dirin et al. [100] used a multi-step method to obtain effective and stable PQDs through a microcarrier-assisted inorganic shell, in which PQDs were first anchored on a salt carrier and then coated with an inorganic shell through the surface-mediated reaction of the precursor. The thermostability could be significantly improved from the obtained powder. Wei et al. [101] integrated $CsPbBr_3$ into CaF_2 NPs, which greatly improved the stability to moisture, light radiation, and anion exchange.

Embedding PQDs into a polymer matrix to obtain stable QD films is one of the most popular technologies, because it does not require expensive vacuum equipment. Shen et al. [102] synthesized polymer-coated $CsPbBr_3$ QDs in situ on a template. The conjugated linoleic acid (CLA) was used as a ligand to passivate the surface defects of QDs, and the CLA crosslinking was triggered under light excitation to form polymer coating and improve the stability of QDs. The FWHM of the prepared composite was narrow, and the PLQY could reach 79.16%. More importantly, the PL intensity could be maintained at 77% after soaking in water for one week due to the protection of the hydrophobic polymer layer. Cai et al. [103] reported an easy synthesis of water-resistant $CsPbBr_3$ PQDs loaded poly(methyl methacrylate) (PMMA) composite microspheres ($CsPbBr_3$@PMMA). Traditional OA was replaced with methacrylic acid (MAA), and the double bonds from methyl methacrylate (MMA) helped $CsPbBr_3$ PQDs be polymerized into PMMA (see Figure 7c). Furthermore, a wide-color-gamut (~129%) white LED was demonstrated by combining the green-emitting $CsPbBr_3$@PMMA composite microspheres and red-emitting K_2SiF_6: Mn^{4+} with a blue LED, which could be used as backlight for liquid crystal displays (LCD).

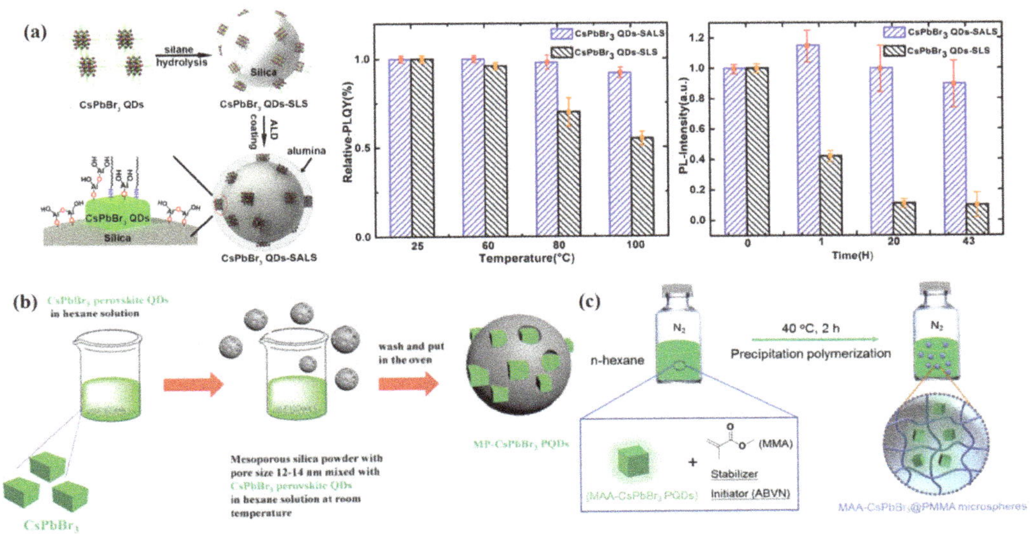

Figure 7. (**a**) Schematic diagram of selective surface passivation of $CsPbBr_3$ QDs silicon luminescent sphere (SALS) by ALD (left); thermal stability, and photostability of as-prepared $CsPbBr_3$ QDs-SLS and QDs-SALS (right). Adapted with permission from [95]. Copyright 2018, American Chemical Society. (**b**) The synthesis process of mesoporous silica green PQDs nanocomposite (MP-PQDs). Reproduced with permission from [96]. Copyright 2016, Wiley. (**c**) Illustration of the preparation process of MAA-$CsPbBr_3$@PMMA microspheres. Reproduced with permission from [103]. Copyright 2019, Wiley.

The glass embedding method is another feasible way to stabilize $CsPbX_3$ QDs during synthesis [104]. The size of the prepared $CsPbX_3$ QDs can be regulated through melt quenching and subsequent thermal treatment, and finally surrounded by a solid glass matrix. These stable $CsPbX_3$ NCs could be easily integrated into display or lighting applications. Hou et al. [105] prepared a composite material based on a metal-organic framework (MOF), zeolitic imidazolate framework (ZIF) glass, and all-inorganic perovskite. It was proved that the interface development process during liquid-phase sintering played an important role in stabilizing the optically active phase of $CsPbX_3$. The rigid, hydrophobic agZIF-62 provided protection for $CsPbI_3$, leading to stable PL emission for $(CsPbI_3)_{0.25}$ $(agZIF\text{-}62)_{0.75}$ after extended (~20 h) sonication in various nonpolar, polar protic, and polar aprotic organic solvents. In addition, the composite also exhibited stability against 10,000 h immersion in water, storage under ambient conditions for 650 days, mild heating, and continuous laser excitation for >5000 s.

Table 3 summarizes the coating of PQDs realized with different encapsulation materials and methods.

Table 3. Luminescence properties of PQDs realized with different packaging materials and methods.

Wrapping	Excitation (nm)	PL (nm)	FWHM (nm)	PLQY (%)	τ (ns)	Stability	Advantages	References
$CsPbBr_3/SiO_2$	350	533	18	–	48.3	73.8% (75% RH, air, 12 h); 36.4% (60 °C, 15 h)	Anion exchange prevention and stability improvement	[90]
$CsPbX_3/ZnS$	365	–	–	70	–	–	More stable, tunable	[91]
$CsPbBr_3/TiO_2$	405	518	32	–	2.1	Stable for 12 weeks (water); ≈75% (UV, 24 h)	Suppress anion exchange and photodegradation	[92]
$CsPbBr_3/Al_2O_3$	365	516	23	65	36.57	PL stable (96 h, water); 80% (450 nm, 200 mW/cm^2, 40 h)	Defect passivation	[95]
$CsPbBr_3$/Mesoporous silica	365	457–698	13–35	–	–	80% (365 nm, 6 W UV, 96 h)	Prevent ion exchange and increase stability	[96]
$MAPbBr_3/NaNO_3$	365	525–526	24	42	155.5	30% (100 °C, 5 h); 80% (365 nm/6 W UV, 14 h)	Improved stability	[99]
$CsPbBr_3$@SiO2/Poly-CLA	365	511	20	79.16	218.11	77% (water, 1 week)	Improved stability	[102]
$CsPbBr_3$@PMMA	395	514	26	32.8	122.2	91% (water, 7 days); stable (water, 1 month)	Improved water resistance and storage stability	[103]
$CsPbI_3$/ZIF glass	–	–	42	>65%	17.6 (average)	80% (water, 10,000 h) no CsPbI3 phases change (air condition) active phase preserved (after 77 K) Over 80% (after 100 °C in air or 80 °C in air for 1000 cycles) 90% (57 mW/cm^2 over 5000 s)	Improved stability	[105]

5. Progress of PQDs in Displays

PQDs has been used for LCD backlights as a semiconductor photoluminescent material. The basic principle is similar to traditional QDs based on their PL characteristics [106]. The white backlight source could be obtained through PQD color conversion and blue LED excitation, and then full-color display can be realized by the color filters inside the LCD panel. PQDs can also be used in a QD color filter (QDCF), which is directly collocated with blue light sources to realize a full-color display. In addition, using green and red QD color conversion layers (QDCCLs) integrated into the display panel is another color realization strategy for self-emissive displays [107–110]. The green and red emission can be obtained with the excitation of the blue OLED or blue LED [111,112]. The display applications based on the PL characteristic of PQDs are shown in Figure 8 for clear comparison. Electrically driven QLED based on electroluminescent (EL) properties will be discussed further.

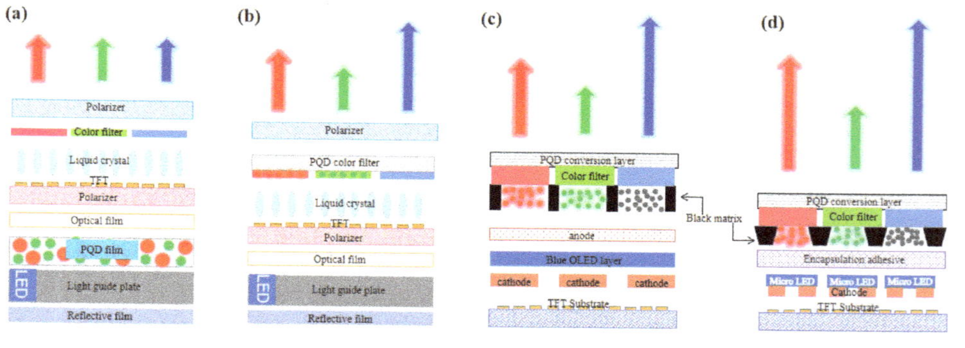

Figure 8. Photoluminescence display application of PQDs: (**a**) QD backlight; (**b**) QDCF; (**c**) full-color OLED with QDCCLs; (**d**) full-color micro-LED with QDCCLs.

5.1. Display Applications Based on PQD Photoluminescence

5.1.1. PQD Backlight

The LCD's color gamut is mainly determined by the backlight and color filter, and the FWHM of the light-emitting material determines the color coordinates of the backlight. In traditional white LED backlights, the spectrum is composed of blue (~450 nm) and yellow light (~580 nm) with a wide spectrum. However, in QD-based backlight, the spectrum is composed of three narrow emission peaks of red, green, and blue (RGB).

Compared with most traditional luminous materials, PQDs have the characteristics of higher efficiency and narrower FWHM, which would become a representative material in displays in the future. Taking the edge-lit LED backlight as an example, the PQD's backlight technology can be divided into four types according to the encapsulation methods.

(1) On chip (chip direct contact). As shown in Figure 9a, PQDs replaces the traditional phosphor materials and are directly encapsulated on the blue LED chips to obtain a white light source [113]. Although this structure has the merits of easy realization and low cost, it requires a highly stable performance of PQDs to ensure efficient luminescence. This is mainly because the PQDs have to be directly in contact with the LED chip. A normally working LED chip will emit dramatically increased heat. The above PQDs have to maintain stability at a temperature even higher than 150°C, which is still a huge challenge to PQDs currently.

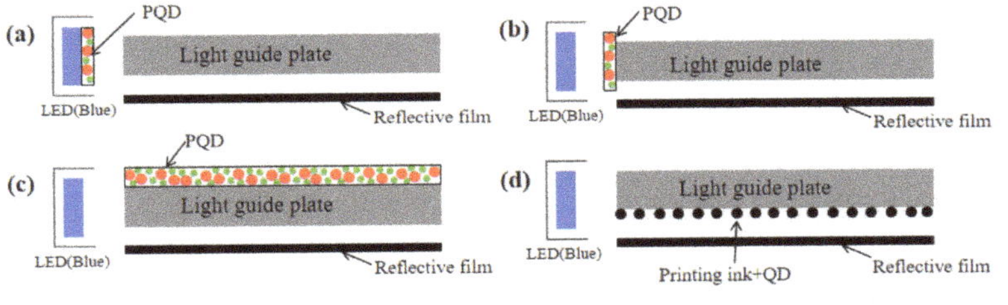

Figure 9. Schematic diagram of PQD backlight structures. (**a**) On chip; (**b**) on edge; (**c**) on surface; (**d**) in dot.

(2) On edge (glass tube encapsulation). As shown in Figure 9b, PQDs are encapsulated into a long glass tube, and placed on the side of a light guide plate with a blue LED light bar. In this structure, PQDs are not in direct contact with a blue LED chip, so the influence of thermal radiation and light radiation from LED chips can be greatly reduced. It has been used for display backlights in 2013, which was called "QD optical tube technology" developed by QD vision in the United States [114]. However, the existing QD glass tube encapsulation technology has the problems of low luminous efficiency (LE) and unfavorable assembly operation, which limits its large-scale application and development. In addition, it is still uncertain whether the environmental isolation level of ordinary glass can meet the need of PQDs.

(3) On surface (optical film or plate integration). As shown in Figure 9c, the PQD film can be placed directly above the light guide plate, and the blue LED is placed on the side of the backlight module. The blue light emitted by the LEDs is shaped into a uniform blue surface light through the joint action of the light guide plate and the bottom reflection film. The blue surface light source then excites PQD film to form a white backlight source. Another on-surface configuration may be the PQD diffuser plate, which should be used for direct-lit mini-LED backlights [115,116]. Because PQDs are far enough from blue LED, they are hardly affected by the thermal radiation and light radiation of the LED chip. Meanwhile, the multi-layer structure also provides a potential to isolate the PQD layer from the environment. The key problem is how to balance the ratio of the three primary colors for a white backlight with a higher stability and a wider color gamut.

(4) In dot (micro-structure integration). As shown in Figure 9d, the micro-structure is fabricated by a screen printing or an inkjet printing process which contains the mixture of red/green QDs and ink, then the mixed QD slurry is printed on the bottom surface of the light guide plate for light mixing and diffusion [94,106,117]. The red and green emissions are converted by red and green QDs under the excitation of the edge-lit blue LED. These micro-structural arrays can be optimized for better uniformity and extraction efficiency [118]. Uniform white light could be achieved by adjusting the proportion of the red and green QDs in the mixture. The current challenge is the ink configuration and environmental exposure.

These photoluminescent PQD devices face their own respective problems. The "on chip" solution is more straightforward and easier to prepare, so it has become the primary choice for researchers. The "on chip" backlight can achieve good white light with color mixing in a limited encapsulation space. In 2015, Zhang and co-workers [25] developed white down-converted (DC) LEDs with a wide color gamut by using HPQDs $MAPbBr_3$ (green-emitting) and rare-earth $K_2SiF_6{:}Mn^{4+}$ phosphors (red-emitting) as color down-converters. This device had CIE coordinates of (0.33, 0.27), 130% NTSC, and an LE of 48 lm/W at 4.9 mA, which could well match the need of display backlights. To further improve the stability, Wang et al. [96] reported an "on chip" structure by introducing mesoporous silica composites. It should be noted that the prepared white-light device demonstrated 113% NTSC after passing through a color filter without the anion exchange issue. In 2022, Wang et al. further showed that the PLQY improvement of full-visible-spectrum IPQDs could be successfully attained by a renewable and low-cost anion exchange resin. The PLQY of three-primary-color IPQDs could be dramatically improved to 93.69%, 89.99%, and 65.03%. Meanwhile, the prepared LED by "On chip" solution provides high brightness and a wide color gamut simultaneously [115]. Although the important display parameters, including the color restoration and stability improvement, have been the focus of attention, the potential of different PQD photoluminescent devices still deserves to be further explored.

5.1.2. PQD Color Conversion Layer

Color definition of LCD is realized through the backlight and color filter in the liquid crystal panel. In addition to backlight, PQDs can also be introduced into a color filter as CCLs, as shown in Figure 8b–d. By employing blue OLED or micro-LED as the excitation source, almost all blue light is converted into desired red or green light by PQDCCLs to realize full-color displays. The blue light should be absorbed as much as possible to reduce eye damage. Therefore, by using PQDs as CCLs, the first consideration is how to maintain the high PLQY while ensuring stability.

Yang et al. [119] compounded PQDs and thermoplastic elastomer into a stretchable and self-healing filter film. Furthermore, they fabricated and sequentially stacked green ($MAPbBr_3$) and red ($CsPbBr_{0.6}I_{2.4}$) composite films, which can be excited by a blue LED to realize white light. It is worth mentioning that PQDs' polymer films with humidity self-healing properties have been further discussed by Cai et al. to support the potential use of polymer conversion films in displays [120]. Yin et al. [121] developed vacuum drying perovskite film preparation technology, and successfully prepared micron-level perovskite films with uniform thickness. The perovskite film with a thickness of 3.8 μm was excited by using 463 nm blue OLED or micro-LED. The brightness of the generated green emission could reach 200 cd/m^2 when the brightness of the excitation source was at 1000 cd/m^2. In addition, only 2% brightness attenuation was observed after 18 days of exposure to the environment. In 2020, Hu et al. [122] used inkjet printing and UV-induced polymerization to obtain micron-thick QD films with uniform surface morphology. By combining QD films with blue OLED or micro-LED displays, green displays were successfully realized via color conversion. Due to the applicability of wet preparation, replacing traditional QDs with PQDs is also feasible, which can support the development of PQDCCLs.

Photoluminescence displays or backlights occupied an important position for the potential application of PQDs. The studies on traditional QD photoluminescence application are increasingly mature these days. The opportunity of PQD's deeper permeation into displays is the way to replace traditional color conversion material and find effective ways to enhance stability.

5.2. Display Applications Based on PQD Electroluminescence

In terms of PQD electroluminescence, quantum-dot light-emitting diodes (QLED) have aroused worldwide concern for decades. They are naturally compatible with attractive flexible or bendable display devices. The booming development of QLED requires the exploration of more efficient emitters, such as PQDs. The characteristics, including high radiative recombination, high defect tolerance, and excellent optical properties, have allowed them to be a rising star as QLED emitters. Numerous studies have been devoted to perovskite QLED (PeQLED) to enhance performance through composition design, surface engineering, and device structure modification [82,123,124].

The first PeQLED was fabricated with $MAPbBr_3$ QDs in 2014 by Schmidt et al. [30]. The demonstrated luminance of this device was extremely low, which was even lower than 1 cd/m^2. Later on, Huang et al. [125] employed size-tunable $MAPbBr_3$ QDs to fabricate QLED, which exhibited improved performance with a maximum brightness of 2503 cd/m^2, current efficiency (CE) of 4.5 cd/A, power efficiency (PE) of 3.5 lm/W, and 1.1% EQE. Xing et al. [126] adopted the device structure of ITO/PEDOT: PSS/$MAPbX_3$ QDs/TPBi/Cs_2CO_3/Al and obtained superior performance with CE of 11.49 cd/A, PE of 7.84 lm/W, and 3.8% EQE. Yan et al. [123] presented an efficient QLED based on $MAPbBr_3$ QDs by achieving charge balance and suppressing the Auger recombination under a low driving voltage. This device showed a maximum luminance of 43,440 cd/m^2, a PE of 30.3 lm/W, and 12.9% EQE.

As an effective strategy, mixing cations have been employed to construct high-efficiency PeQLED. Cho et al. [127] first reported the synthesis of $MA_{1-x}Cs_xPbBr_3$ QDs and the corresponding $MA_{0.7}Cs_{0.3}PbBr_3$-based QLED, which had a maximum luminance of 24,510 cd/m^2, a CE of 4.1 cd/A, and 1.3% EQE. Zhang [128] exploited the mixed cation $FA_{0.8}Cs_{0.2}PbBr_3$ to fabricate QLED, which achieved a high luminance of 55,005 cd/m^2, a CE of 10.09 cd/A, and 2.80% EQE. Red $FA_{0.87}Cs_{0.13}PbI_3$ NCs with γ-butyrolactone was utilized as a solvent to realize a QLED with a maximum luminance of 218 cd/m^2 and a peak EQE of 15.8% in 2019 [129]. Pan et al. [130] incorporated Ni^{2+} ions into $CsPbCl_xBr_{3-x}$ QDs by a supersaturated recrystallization synthetic method and modulated the Cl/Br element ratios to obtain an efficient blue PeQLED. The corresponding device presented a maximum luminance of 612 cd/m^2 and 2.4% EQE (Figure 10).

Actually, surface ligand engineering plays an important role in the performance enhancement of PeQLED. Although surface capping ligands could remove surface trap states to improve the stability of PQDs, they also form an insulating film to hinder the charge injection and transport inside the light-emitting layer of PeQLED. Therefore, appropriate surface ligand modification benefits the performance enhancement of PeQLED. Lee et al. [131] utilized a short amine ligand to enhance the efficiency of $FAPbBr_3$-based PeQLED. The reported $FAPbBr_{3-n}$-butylamine-based PeQLED had a high performance with CE of 9.16 cd/A, a PE of 6.4 lm/W, and 2.5% EQE. Han [124] exploited 3,3-diphenylpropylamine bromide (DPPA-Br) as surface capping ligands to fabricate uniform $FAPbBr_3$ QDs. The resulting PeQLED demonstrated an excellent performance with a maximum luminance of 13,970 cd/m^2, a CE of 66.3 cd/A, and 16.3% EQE. Song et al. [82] explored an organic-inorganic hybrid ligand to passivate the surface trap state and enhance the carrier injection of PQDs, including the use of $ZnBr_2$, $MnBr_2$, $GaBr_3$, and $InBr_3$. As a result, the $ZnBr_2$-passivated PeQLED showed superior performance with a maximum luminance of 76,940 cd/m^2, a CE of 66.7 cd/A, a PE of 65.9 lm/W, and 16.48% EQE, while the $MnBr_2$-passivated PeQLED presented a maximum luminance of 100,080 cd/m^2, a CE of 60.6 cd/A, and 15.6% EQE (Figure 11). In 2021, Li et al. [132] used FABr to compensate the surface

Bromine vacancy of inorganic cesium lead halide PQDs. The corresponding PeQLED with FABr-treated $CsPbBr_3$ QDs (green-emitting) achieved a promising EQE of 7.94% and a luminance of 14,790 cd/m^2, which was much higher than those of the pristine (1.78%, 4640 cd/m^2) and the DDAB-treated samples (3.91%, 167 cd/m^2). This methodology is also suitable for the surface defect passivation of blue-emitting $CsPbBr_{1.3}Cl_{1.7}$ and red-emitting $CsPbBrI_2$ QDs. The results clearly highlighted that heterogeneous post-passivation could boost the PL and EL performance of inorganic cesium lead halide PQDs simultaneously.

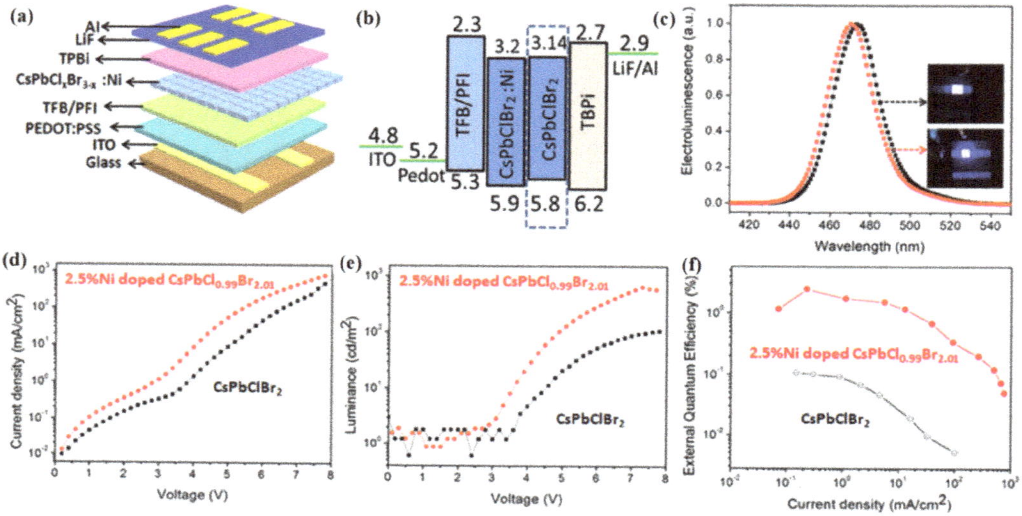

Figure 10. (**a**,**b**) Schematic diagram and band level of the PeQLED based on 2.5% Ni^{2+} ion-doped $CsPbCl_{0.99}Br_{2.01}$ QDs. (**c**) The EL spectra and photos of the PeQLED based on $CsPbClBr_2$ QDs (black) and 2.5% Ni^{2+}—doped $CsPbCl_{0.99}Br_{2.01}$ QDs (red), respectively. (**d**–**f**) Current density, luminance, and EQE of the corresponding devices. Adapted with permission from [130]. Copyright 2020, American Chemical Society.

Surface ligand passivation can improve the PLQY of QD, but their insulating nature impedes the charge injection in a QD film and further negatively impacts the PeQLED's performance. Hence, appropriate ligand choice and effective ligand exchange play important roles in building a high-efficiency PeQLED. Moreover, optimizing device inter-layer structures with matched energy levels to increase radiative recombination efficiency is another major method in the achievement of efficient PeQLED. Recently, Khan et al. [133] optimized the device architecture and energy level matching of $CsPbBr_3$-based PeQLED through the introduction of Li-doped TiO_2 nanoparticles as an electron transport layer (ETL). Compared with the untreated device (3 V turn-on voltage and 5.6 cd/A CE), the resulting device exhibited 2V turn-on voltage and an enhanced CE of 15.2 cd/A with the Li-doped TiO_2 ETL.

Other effective measures are proposed to improve PeQLED. Dong et al. [134] reported PQDs resurfacing to achieve a bipolar shell consisting of an inner anion shell, and an outer shell comprised of cations and polar solvent molecules. The outer shell was electrostatically adsorbed to the negatively charged inner shell. This approach produced strongly confined PQD solids with improved carrier mobility ($\geq$0.01 cm^2 V^{-1} s^{-1}) and reduced trap density. Efficient blue and green PeLEDs can be fabricated by exploiting $CsPbBr_3$ QDs with reduced trap density and improved mobility. Tsai et al. [135] demonstrated that perovskite NCs stabilized in the MOF thin films could maintain decent PL and EL against continuous ultraviolet irradiation, heat, and electrical stress. Bright and stable LEDs were demonstrated with a maximum EQE of over 15% and a high brightness of over 105 cd/m^2

after stabilization. During PeLED operation, the nanocrystals can be well preserved, free of ion migration or crystal merging through protection by the MOF matrix, leading to a stable performance over 50 h.

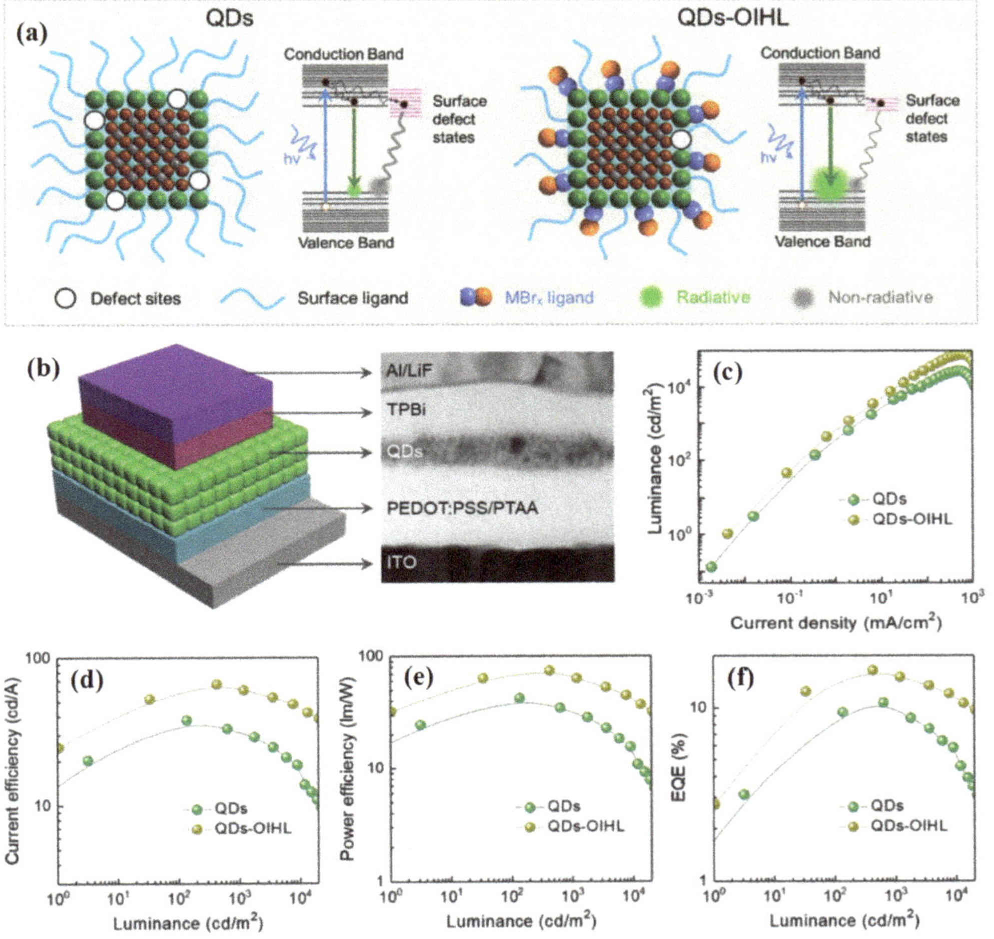

Figure 11. (**a**) Schematic of radiative/nonradiative recombination of PQDs with and without OIHL. (**b**) Illustration of multilayer PeQLED device and cross-sectional TEM image. (**c–f**) Luminance, CE, PE, and EQE of the devices. Reproduced with permission from [82]. Copyright 2018, Wiley.

Figure 12 summarizes the development of EQE of full-color PeQLED devices. Up to now, the maximum EQE values of red, green, and blue QLED have reached 21.6%, 23.4%, and 12.3%, respectively, which has increased by ~20% since 2015. However, so far, the PeQLED's lifetime is still far lower than the commercial standard of 10,000 h. The main obstacle to the commercialization of PeQLED is its instability in the environment. Consequently, improving the performance and stability of PQDs is still the main goal for future development.

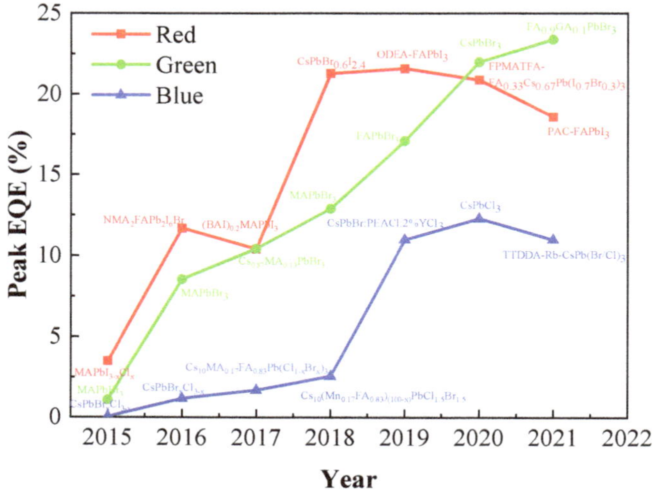

Figure 12. Evolution curve of the peak EQE of perovskite QLED.

6. Conclusions

In this paper, we summarize the synthesis methods of PQDs and analyze the effects of ion doping, ligand modification, and coating engineering on PQDs. Furthermore, the applications of PQDs in photoluminescence and electroluminescence displays are described in detail. It is believed that the following aspects require more attention as regards the display future: (1) large-area and pixelated preparation of PQDs with excellent homogeneity; (2) use of ligand modification, ion doping, or coating strategies to synthetically improve the environmental stability of PQD while retaining high PLQY; (3) development of physical and chemical properties to realize more diverse PQD display devices, such as flexible and bendable displays; (4) lead-free PQDs with excellent luminescence efficiency and high environmental friendliness to meet the needs of sustainable development.

Author Contributions: X.R. and X.Z.: conceptualization, writing—review and editing; H.X., J.C. and C.W.: review and suggestions; E.C.: funding acquisition, review and editing. S.X., Y.Y., J.S., Q.Y. and T.G.: review and funding acquisition. All authors have read and agreed to the published version of the manuscript.

Funding: This research was funded by National Natural Science Foundation of China (Grant No. 62175032); the Fujian Provincial Natural Science Foundation project (Grant No. 2021J01579); the Fujian Provincial key science and technology project (2021HZ021001); and the Fujian Science & Technology Innovation Laboratory for Optoelectronic Information of China (2020ZZ111, 2020ZZ112).

Institutional Review Board Statement: Not applicable.

Informed Consent Statement: Not applicable.

Data Availability Statement: Not applicable.

Conflicts of Interest: There are no conflict of interest.

References

1. Su, Q.; Chen, S. Thermal assisted up-conversion electroluminescence in quantum dot light emitting diodes. *Nat. Commun.* **2022**, *13*, 369. [CrossRef]
2. Yang, L.; Li, X.; Yang, Q.; Wang, S.; Tian, H.; Ding, J.; Wang, L. Efficient narrowband red electroluminescence from a thermally activated delayed fluorescence polymer and quantum dot hybrid. *Chem. Eng. J.* **2022**, *436*, 135221. [CrossRef]
3. Singh, R.K.; Chen, L.H.; Singh, A.; Jain, N.; Singh, J.; Lu, C.H. Progress of backlight devices: Emergence of halide perovskite quantum Dots/Nanomaterials. *Front. Nanotechnol.* **2022**, *29*, 863312. [CrossRef]

4. Bornacelli, J.; Torres-Torres, C.; Silva-Pereyra, H.G.; Labrada-Delgado, G.J.; Crespo-Sosa, A.; Cheang-Wong, J.C.; Oliver, A. Superlinear photoluminescence by ultrafast laser pulses in dielectric matrices with metal nanoclusters. *Sci. Rep.* **2019**, *9*, 5699. [CrossRef]
5. Xu, Y.; Xu, R.; Wang, Z.; Zhou, Y.; Shen, Q.; Ji, W.; Dang, D.; Meng, L.; Tang, B.Z. Recent advances in luminescent materials for super-resolution imaging via stimulated emission depletion nanoscopy. *Chem. Soc. Rev.* **2021**, *50*, 667–690. [CrossRef]
6. Wang, Q.; Tong, Y.; Yang, M.; Ye, H.; Liang, X.; Wang, X.; Xiang, W. ZnO induced self-crystallization of CsPb(Br/I)3 nanocrystal glasses with improved stability for backlight display application. *J. Mater. Sci. Technol.* **2022**, *121*, 140–147. [CrossRef]
7. Kang, C.; Tao, S.; Yang, F.; Yang, B. Aggregation and luminescence in carbonized polymer dots. *Aggregate* **2022**, *3*, e169. [CrossRef]
8. Li, K.; Shen, C. White LED based on nano-YAG:Ce^{3+}/YAG:Ce^{3+},Gd^{3+} hybrid phosphors. *Optik* **2012**, *123*, 621–623. [CrossRef]
9. Chen, H.; He, J.; Wu, S.T. Recent Advances on Quantum-Dot-Enhanced Liquid-Crystal Displays. *IEEE J. Sel. Top. Quant.* **2017**, *23*, 1–11. [CrossRef]
10. Tsai, Y.; Nguyen, H.; Lazarowska, A.; Mahlik, S.; Grinberg, M.; Liu, R. Improvement of the water resistance of a Narrow-Band Red-Emitting $SrLiAl_3N_4$:eu^{2+} phosphor synthesized under high isostatic pressure through coating with an organosilica layer. *Angew. Chem. Int. Ed.* **2016**, *55*, 9652–9656. [CrossRef]
11. Wu, W.; Fang, M.; Zhou, W.; Lesniewski, T.; Mahlik, S.; Grinberg, M.; Brik, M.G.; Sheu, H.; Cheng, B.; Wang, J.; et al. High color rendering index of Rb_2GeF_6:mn^{4+} for Light-Emitting diodes. *Chem. Mater.* **2017**, *29*, 935–939. [CrossRef]
12. Fang, M.; Wu, W.; Jin, Y.; Lesniewski, T.; Mahlik, S.; Grinberg, M.; Brik, M.G.; Srivastava, A.M.; Chiang, C.; Zhou, W.; et al. Control of luminescence by tuning of crystal symmetry and local structure in mn^{4+}—Activated narrow band fluoride phosphors. *Angew. Chem. Int. Ed.* **2018**, *57*, 1797–1801. [CrossRef] [PubMed]
13. Zhao, B.; Wang, Q.; Li, D.; Yang, H.; Bai, X.; Li, S.; Liu, P.; Sun, X. Red and green quantum dot color filter for Full-Color Micro-LED arrays. *Micromachines* **2022**, *13*, 595. [CrossRef] [PubMed]
14. Zhao, C.; Li, Y.; Ye, W.; Shen, X.; Wen, Z.; Yuan, X.; Cao, Y.; Ma, C. Ligand-Free CsPbBr3 perovskite quantum dots in Silica-Aerogel composites with enhanced stability for w-LED and display by substituting pb^{2+} with pr^{3+} or gd^{3+} ions. *Adv. Opt. Mater.* **2022**, *10*, 2102200. [CrossRef]
15. Yang, Z.; Gao, M.; Wu, W.; Yang, X.; Sun, X.W.; Zhang, J.; Wang, H.; Liu, R.; Han, C.; Yang, H.; et al. Recent advances in quantum dot-based light-emitting devices: Challenges and possible solutions. *Mater. Today* **2019**, *24*, 69–93. [CrossRef]
16. Hanifi, D.A.; Bronstein, N.D.; Koscher, B.A.; Nett, Z.; Swabeck, J.K.; Takano, K.; Schwartzberg, A.M.; Maserati, L.; Vandewal, K.; van de Burgt, Y.; et al. Redefining near-unity luminescence in quantum dots with photothermal threshold quantum yield. *Science* **2019**, *363*, 1199–1202. [CrossRef]
17. Chen, O.; Zhao, J.; Chauhan, V.P.; Cui, J.; Wong, C.; Harris, D.K.; Wei, H.; Han, H.; Fukumura, D.; Jain, R.K.; et al. Compact high-quality CdSe-CdS core-shell nanocrystals with narrow emission linewidths and suppressed blinking. *Nat. Mater.* **2013**, *12*, 445–451. [CrossRef]
18. Won, Y.; Cho, O.; Kim, T.; Chung, D.; Kim, T.; Chung, H.; Jang, H.; Lee, J.; Kim, D.; Jang, E. Highly efficient and stable InP/ZnSe/ZnS quantum dot light-emitting diodes. *Nature* **2019**, *575*, 634–638. [CrossRef]
19. Kim, Y.; Ham, S.; Jang, H.; Min, J.H.; Chung, H.; Lee, J.; Kim, D.; Jang, E. Bright and uniform green light emitting InP/ZnSe/ZnS quantum dots for wide color gamut displays. *ACS Appl. Nano Mater.* **2019**, *2*, 1496–1504. [CrossRef]
20. Swarnkar, A.; Chulliyil, R.; Ravi, V.K.; Irfanullah, M.; Chowdhury, A.; Nag, A. Colloidal CsPbBr3 Perovskite Nanocrystals: Luminescence beyond Traditional Quantum Dots. *Angew. Chem. Int. Ed.* **2015**, *54*, 15424–15428. [CrossRef]
21. Lv, W.; Li, L.; Xu, M.; Hong, J.; Tang, X.; Xu, L.; Wu, Y.; Zhu, R.; Chen, R.; Huang, W. Improving the stability of metal halide perovskite quantum dots by encapsulation. *Adv. Mater.* **2019**, *31*, 1900682. [CrossRef] [PubMed]
22. Wang, X.; Bao, Z.; Chang, Y.; Liu, R. Perovskite quantum dots for application in high color gamut backlighting display of Light-Emitting diodes. *ACS Energy Lett.* **2020**, *5*, 3374–3396. [CrossRef]
23. Abdi-Jalebi, M.; Pazoki, M.; Philippe, B.; Dar, M.I.; Alsari, M.; Sadhanala, A.; Diyitini, G.; Imani, R.; Lilliu, S.; Kullgren, J.; et al. Dedoping of lead halide perovskites incorporating monovalent cations. *ACS Nano* **2018**, *12*, 7301–7311. [CrossRef] [PubMed]
24. Eperon, G.E.; Stranks, S.D.; Menelaou, C.; Johnston, M.B.; Herz, L.M.; Snaith, H.J. Formamidinium lead trihalide: A broadly tunable perovskite for efficient planar heterojunction solar cells. *Energy Environ. Sci.* **2014**, *7*, 982–988. [CrossRef]
25. Zhang, F.; Zhong, H.; Chen, C.; Wu, X.; Hu, X.; Huang, H.; Han, J.; Zou, B.; Dong, Y. Brightly luminescent and color-tunable colloidal $CH_3NH_3PbX_3$ (X = Br, I, Cl) quantum dots: Potential alternatives for display technology. *ACS Nano* **2015**, *9*, 4533–4542. [CrossRef]
26. Protesescu, L.; Yakunin, S.; Bodnarchuk, M.I.; Krieg, F.; Caputo, R.; Hendon, C.H.; Yang, R.X.; Walsh, A.; Kovalenko, M.V. Nanocrystals of cesium lead halide perovskites ($CsPbX_3$, X = Cl, Br, and I): Novel optoelectronic materials showing bright emission with wide color gamut. *Nano Lett.* **2015**, *15*, 3692–3696. [CrossRef]
27. Sichert, J.A.; Tong, Y.; Mutz, N.; Vollmer, M.; Fischer, S.; Milowska, K.Z.; García Cortadella, R.; Nickel, B.; Cardenas-Daw, C.; Stolarczyk, J.K.; et al. Quantum size effect in organometal halide perovskite nanoplatelets. *Nano Lett.* **2015**, *15*, 6521–6527. [CrossRef]
28. Gonzalez-Carrero, S.; Galian, R.E.; Perez-Prieto, J. Maximizing the emissive properties of $CH_3NH_3PbBr_3$ perovskite nanoparticles. *J. Mater. Chem. A* **2015**, *3*, 9187–9193. [CrossRef]
29. Huang, H.; Susha, A.S.; Kershaw, S.V.; Hung, T.F.; Rogach, A.L. Control of emission color of high quantum yield $CH_3NH_3PbBr_3$ perovskite quantum dots by precipitation temperature. *Adv. Sci.* **2015**, *2*, 1500194. [CrossRef]

30. Schmidt, L.C.; Pertegas, A.; Gonzalez-Carrero, S.; Malinkiewicz, O.; Agouram, S.; Minguez Espallargas, G.; Bolink, H.J.; Galian, R.E.; Perez-Prieto, J. Nontemplate synthesis of $CH_3NH_3PbBr_3$ perovskite nanoparticles. *J. Am. Chem. Soc.* **2014**, *136*, 850–853. [CrossRef]
31. D'Innocenzo, V.; Grancini, G.; Alcocer, M.J.P.; Kandada, A.R.S.; Stranks, S.D.; Lee, M.M.; Lanzani, G.; Snaith, H.J.; Petrozza, A. Excitons versus free charges in organo-lead tri-halide perovskites. *Nat. Commun.* **2014**, *5*, 3586. [CrossRef] [PubMed]
32. Sun, S.; Salim, T.; Mathews, N.; Duchamp, M.; Boothroyd, C.; Xing, G.; Sum, T.C.; Lam, Y.M. The origin of high efficiency in low-temperature solution-processable bilayer organometal halide hybrid solar cells. *Energy Environ. Sci.* **2014**, *7*, 399–407. [CrossRef]
33. Manser, J.S.; Christians, J.A.; Kamat, P.V. Intriguing Optoelectronic Properties of Metal Halide Perovskites. *Chem. Rev.* **2016**, *116*, 12956–13008. [CrossRef]
34. Xing, G.; Mathews, N.; Sun, S.; Lim, S.S.; Lam, Y.M.; Graetzel, M.; Mhaisalkar, S.; Sum, T.C. Long-Range balanced electron- and Hole-Transport lengths in Organic-Inorganic $CH_3NH_3PbI_3$. *Science* **2013**, *342*, 344–347. [CrossRef]
35. Motta, C.; El-Mellouhi, F.; Kais, S.; Tabet, N.; Alharbi, F.; Sanvito, S. Revealing the role of organic cations in hybrid halide perovskite $CH_3NH_3PbI_3$. *Nat. Commun.* **2015**, *6*, 7026. [CrossRef]
36. Jeon, N.J.; Noh, J.H.; Kim, Y.C.; Yang, W.S.; Ryu, S.; Seok, S.I. Solvent engineering for high-performance inorganic-organic hybrid perovskite solar cells. *Nat. Mater.* **2014**, *13*, 897–903. [CrossRef]
37. Jiang, Q.; Zhang, L.; Wang, H.; Yang, X.; Meng, J.; Liu, H.; Yin, Z.; Wu, J.; Zhang, X.; You, J. Enhanced electron extraction using SnO_2 for high-efficiency planar-structure $HC(NH_2)_2PbI_3$-based perovskite solar cells. *Nat. Energy* **2017**, *2*, 16177. [CrossRef]
38. Ponseca, C.S., Jr.; Savenije, T.J.; Abdellah, M.; Zheng, K.; Yartsev, A.; Pascher, T.; Harlang, T.; Chabera, P.; Pullerits, T.; Stepanov, A.; et al. Organometal halide perovskite solar cell materials rationalized: Ultrafast charge generation, high and Microsecond-Long balanced mobilities, and slow recombination. *J. Am. Chem. Soc.* **2014**, *136*, 5189–5192. [CrossRef]
39. Zhang, D.W. *Preparation, Properties and Application of Perovskite Quantum Dots and Metal—Organic Framework Composites*; University of Science & Technology Beijing: Beijing, China, 2019.
40. Bai, Y.; Hao, M.; Ding, S.; Chen, P.; Wang, L. Surface Chemistry Engineering of Perovskite Quantum Dots: Strategies, Applications, and Perspectives. *Adv. Mater.* **2022**, *34*, 2105958. [CrossRef]
41. Byun, J.; Cho, H.; Wolf, C.; Jang, M.; Sadhanala, A.; Friend, R.H.; Yang, H.; Lee, T.-W. Efficient Visible Quasi-2D Perovskite Light-Emitting Diodes. *Adv. Mater.* **2016**, *28*, 7515–7520. [CrossRef]
42. Bekenstein, Y.; Koscher, B.A.; Eaton, S.W.; Yang, P.; Alivisatos, A.P. Highly Luminescent Colloidal Nanoplates of Perovskite Cesium Lead Halide and Their Oriented Assemblies. *J. Am. Chem. Soc.* **2015**, *137*, 16008–16011. [CrossRef] [PubMed]
43. Zhou, H.; Yuan, S.; Wang, X.; Xu, T.; Wang, X.; Li, H.; Zheng, W.; Fan, P.; Li, Y.; Sun, L.; et al. Vapor Growth and Tunable Lasing of Band Gap Engineered Cesium Lead Halide Perovskite Micro/Nanorods with Triangular Cross Section. *ACS Nano* **2017**, *11*, 1189–1195. [CrossRef] [PubMed]
44. Wong, A.B.; Lai, M.; Eaton, S.W.; Yu, Y.; Lin, E.; Dou, L.; Fu, A.; Yang, P. Growth and Anion Exchange Conversion of $CH_3NH_3PbX_3$ Nanorod Arrays for Light-Emitting Diodes. *Nano Lett.* **2015**, *15*, 5519–5524. [CrossRef] [PubMed]
45. Zhu, H.; Fu, Y.; Meng, F.; Wu, X.; Gong, Z.; Ding, Q.; Gustafsson, M.V.; Trinh, M.T.; Jin, S.; Zhu, X.Y. Lead halide perovskite nanowire lasers with low lasing thresholds and high quality factors. *Nat. Mater.* **2015**, *14*, 636–642. [CrossRef]
46. Alivisatos, A.P. Semiconductor clusters, nanocrystals, and quantum dots. *Science* **1996**, *271*, 933–937. [CrossRef]
47. Nedelcu, G.; Protesescu, L.; Yakunin, S.; Bodnarchuk, M.I.; Grotevent, M.J.; Kovalenko, M.V. Fast Anion-Exchange in Highly Luminescent Nanocrystals of Cesium Lead Halide Perovskites (CsPbX3, X = Cl, Br, I). *Nano Lett.* **2015**, *15*, 5635–5640. [CrossRef]
48. Li, X.; Wu, Y.; Zhang, S.; Cai, B.; Gu, Y.; Song, J.; Zeng, H. $CsPbX_3$ Quantum Dots for Lighting and Displays: Room-Temperature Synthesis, Photoluminescence Superiorities, Underlying Origins and White Light-Emitting Diodes. *Adv. Funct. Mater.* **2016**, *26*, 2435–2445. [CrossRef]
49. Tong, Y.; Bladt, E.; Aygüler, M.F.; Manzi, A.; Milowska, K.Z.; Hintermayr, V.A.; Docampo, P.; Bals, S.; Urban, A.S.; Polavarapu, L.; et al. Highly luminescent cesium lead halide perovskite nanocrystals with tunable composition and thickness by ultrasonication. *Angew. Chem. Int. Ed.* **2016**, *55*, 13887–13892. [CrossRef]
50. Pan, Q.; Hu, H.; Zou, Y.; Chen, M.; Wu, L.; Yang, D.; Yuan, X.; Fan, J.; Sun, B.; Zhang, Q. Microwave-assisted synthesis of high-quality "all-inorganic" $CsPbX_3$ (X = Cl, Br, I) perovskite nanocrystals and their application in light emitting diodes. *J. Mater. Chem. C* **2017**, *5*, 10947–10954. [CrossRef]
51. Chen, M.; Zou, Y.; Wu, L.; Pan, Q.; Yang, D.; Hu, H.; Tan, Y.; Zhong, Q.; Xu, Y.; Liu, H.; et al. Solvothermal Synthesis of High-Quality All-Inorganic Cesium Lead Halide Perovskite Nanocrystals: From Nanocube to Ultrathin Nanowire. *Adv. Funct. Mater.* **2017**, *27*, 1701121. [CrossRef]
52. Jana, A.; Mittal, M.; Singla, A.; Sapra, S. Solvent-free, mechanochemical syntheses of bulk trihalide perovskites and their nanoparticles. *Chem. Commun.* **2017**, *53*, 3046–3049. [CrossRef] [PubMed]
53. Protesescu, L.; Yakunin, S.; Nazarenko, O.; Dirin, D.N.; Kovalenko, M.V. Low-Cost Synthesis of Highly Luminescent Colloidal Lead Halide Perovskite Nanocrystals by Wet Ball Milling. *ACS Appl. Nano Mater.* **2018**, *1*, 1300–1308. [CrossRef] [PubMed]
54. Palazon, F.; El Ajjouri, Y.; Sebastia-Luna, P.; Lauciello, S.; Manna, L.; Bolink, H.J. Mechanochemical synthesis of inorganic halide perovskites: Evolution of phase-purity, morphology, and photoluminescence. *J. Mater. Chem. C* **2019**, *7*, 11406–11410. [CrossRef]

55. Guo, P.; Hossain, M.K.; Shen, X.; Sun, H.; Yang, W.; Liu, C.; Ho, C.Y.; Kwok, C.K.; Tsang, S.; Luo, Y.; et al. Room-temperature red-green-blue whispering-gallery mode lasing and white-light emission from cesium lead halide perovskite ($CsPbX_3$, X = Cl, Br, I) microstructures. *Adv. Opt. Mater.* **2018**, *6*, 1700993. [CrossRef]
56. Lignos, I.; Maceiczyk, R.; de Mello, A.J. Microfluidic Technology: Uncovering the Mechanisms of Nanocrystal Nucleation and Growth. *Acc. Chem. Res.* **2017**, *50*, 1248–1257. [CrossRef]
57. Lignos, I.; Morad, V.; Shynkarenko, Y.; Bernasconi, C.; Maceiczyk, R.M.; Protesescu, L.; Bertolotti, F.; Kumar, S.; Ochsenbein, S.T.; Masciocchi, N.; et al. Exploration of Near-Infrared-Emissive Colloidal Multinary Lead Halide Perovskite Nanocrystals Using an Automated Microfluidic Platform. *ACS Nano* **2018**, *12*, 5504–5517. [CrossRef]
58. Li, X.; Cao, F.; Yu, D.; Chen, J.; Sun, Z.; Shen, Y.; Zhu, Y.; Wang, L.; Wei, Y.; Wu, Y.; et al. All inorganic halide perovskites nanosystem: Synthesis, structural features, optical properties and optoelectronic applications. *Small* **2017**, *13*, 1603996. [CrossRef]
59. Wang, J.; Liu, X.; Zhou, L.; Shen, W.; Li, M.; He, R. Highly luminescent and stable quasi-2D perovskite quantum dots by introducing large organic cations. *Nanoscale Adv.* **2021**, *3*, 5393–5398. [CrossRef]
60. Jiang, J.; Liu, F.; Shen, Q.; Tao, S. The role of sodium in stabilizing tin-lead (Sn-Pb) alloyed perovskite quantum dots. *J. Mater. Chem. A* **2021**, *9*, 12087–12098. [CrossRef]
61. Liu, Y.; Pan, G.; Wang, R.; Shao, H.; Wang, H.; Xu, W.; Cui, H.; Song, H. Considerably enhanced exciton emission of $CsPbCl_3$ perovskite quantum dots by the introduction of potassium and lanthanide ions. *Nanoscale* **2018**, *10*, 14067–14072. [CrossRef]
62. Huang, S.; Wang, B.; Zhang, Q.; Li, Z.; Shan, A.; Li, L. Postsynthesis Potassium-Modification method to improve stability of $CsPbBr_3$ perovskite nanocrystals. *Adv. Opt. Mater.* **2018**, *6*, 1701106. [CrossRef]
63. Baek, S.; Kim, S.; Noh, J.Y.; Heo, J.H.; Im, S.H.; Hong, K.-H.; Kim, S.-W. Development of Mixed-Cation $Cs_xRb_{1-x}PbX_3$ Perovskite Quantum Dots and Their Full-Color Film with High Stability and Wide Color Gamut. *Adv. Opt. Mater.* **2018**, *6*, 1800295. [CrossRef]
64. Shao, H.; Bai, X.; Cui, H.; Pan, G.; Jing, P.; Qu, S.; Zhu, J.; Zhai, Y.; Dong, B.; Song, H. White light emission in Bi^{3+}/Mn^{2+} ion co-doped $CsPbCl_3$ perovskite nanocrystals. *Nanoscale* **2018**, *10*, 1023–1029. [CrossRef] [PubMed]
65. Shen, Y.; Yin, J.; Cai, B.; Wang, Z.; Dong, Y.; Xu, X.; Zeng, H. Lead-free, stable, high-efficiency (52%) blue luminescent $FA_3Bi_2Br_9$ perovskite quantum dots. *Nanoscale Horiz.* **2020**, *5*, 580–585. [PubMed]
66. Luo, C.; Li, W.; Fu, J.; Yang, W. Constructing gradient energy levels to promote exciton energy transfer for photoluminescence controllability of All-Inorganic perovskites and application in Single-Component WLEDs. *Chem. Mater.* **2019**, *31*, 5616–5624.
67. Bi, C.; Wang, S.; Li, Q.; Kershaw, S.V.; Tian, J.; Rogach, A.L. Thermally stable Copper(II)-Doped cesium lead halide perovskite quantum dots with strong blue emission. *J. Phys. Chem. Lett.* **2019**, *10*, 943–952. [CrossRef]
68. Xu, H.; Liang, J.; Zhang, Z.; Deng, Z.; Qiu, Y.; He, M.; Wang, J.; Yang, Y.; Chen, C. Lead-free bright blue light-emitting cesium halide nanocrystals by zinc doping. *RSC Adv.* **2021**, *11*, 2437–2445. [CrossRef]
69. Hu, Y.; Zhang, X.; Yang, C.; Ji, L.; Wang, L. Fe^{2+} doped in $CsPbCl_3$ perovskite nanocrystals: Impact on the luminescence and magnetic properties. *RSC Adv.* **2019**, *9*, 33017–33022.
70. Zou, S.; Yang, G.; Yang, T.; Zhao, D.; Gan, Z.; Chen, W.; Zhong, H.; Wen, X.; Jia, B.; Zou, B. Template-Free synthesis of High-Yield Fe-Doped cesium lead halide perovskite ultralong microwires with enhanced Two-Photon absorption. *J. Phys. Chem. Lett.* **2018**, *9*, 4878–4885.
71. Chen, D.; Zhou, S.; Tian, F.; Ke, H.; Jiang, N.; Wang, S.; Peng, Y.; Liu, Y. Halogen-hot-injection synthesis of Mn-doped $CsPb(Cl/Br)_3$ nanocrystals with blue/orange dual-color luminescence and high photoluminescence quantum yield. *Adv. Opt. Mater.* **2019**, *7*, 1901082. [CrossRef]
72. Yang, H.; Yin, W.; Dong, W.; Gao, L.; Tan, C.; Li, W.; Zhang, X.; Zhang, J. Enhancing light-emitting performance and stability in $CsPbBr_3$ perovskite quantum dots via simultaneous doping and surface passivation. *J. Mater. Chem. C* **2020**, *8*, 14439–14445. [CrossRef]
73. Gualdrón-Reyes, A.F.; Macias-Pinilla, D.F.; Masi, S.; Echeverría-Arrondo, C.; Agouram, S.; Muñoz-Sanjosé, V.; Rodríguez-Pereira, J.; Macak, J.M.; Mora-Seró, I. Engineering Sr-doping for enabling long-term stable $FAPb_{1-x}Sr_xI_3$ quantum dots with 100% photoluminescence quantum yield. *J. Mater. Chem. C* **2021**, *9*, 1555–1566. [CrossRef]
74. Sun, M.-J.; Zheng, C.; Gao, Y.; Johnston, A.; Najarian, A.M.; Wang, P.-X.; Voznyy, O.; Hoogland, S.; Sargent, E.H. Linear Electro-Optic Modulation in Highly Polarizable Organic Perovskites. *Adv. Mater.* **2021**, *33*, 2006368. [CrossRef] [PubMed]
75. Zhong, Q.; Cao, M.; Xu, Y.; Li, P.; Zhang, Y.; Hu, H.; Yang, D.; Xu, Y.; Wang, L.; Li, Y.; et al. L-Type Ligand-Assisted Acid-Free Synthesis of $CsPbBr_3$ Nanocrystals with Near-Unity Photoluminescence Quantum Yield and High Stability. *Nano Lett.* **2019**, *19*, 4151–4157. [CrossRef] [PubMed]
76. Song, J.; Li, J.; Xu, L.; Li, J.; Zhang, F.; Han, B.; Shan, Q.; Zeng, H. Room-Temperature Triple-Ligand Surface Engineering Synergistically Boosts Ink Stability, Recombination Dynamics, and Charge Injection toward EQE-11.6% Perovskite QLEDs. *Adv. Mater.* **2018**, *30*, 1800764. [CrossRef] [PubMed]
77. Bao, Z.; Wang, W.; Tsai, H.-Y.; Wang, H.-C.; Chen, S.; Liu, R.-S. Photo-/electro-luminescence enhancement of $CsPbX_3$ (X = Cl, Br, or I) perovskite quantum dots via thiocyanate surface modification. *J. Mater. Chem. C* **2020**, *8*, 1065–1071. [CrossRef]
78. Park, J.H.; Lee, A.-y.; Yu, J.C.; Nam, Y.S.; Choi, Y.; Park, J.; Song, M.H. Surface Ligand Engineering for Efficient Perovskite Nanocrystal-Based Light-Emitting Diodes. *ACS Appl. Mater.* **2019**, *11*, 8428–8435. [CrossRef]
79. Huang, Y.; Luan, W.; Liu, M.; Turyanska, L. DDAB-assisted synthesis of iodine-rich $CsPbI_3$ perovskite nanocrystals with improved stability in multiple environments. *J. Mater. Chem. C* **2020**, *8*, 2381–2387. [CrossRef]

80. Li, X.; Cai, W.; Guan, H.; Zhao, S.; Cao, S.; Chen, C.; Liu, M.; Zang, Z. Highly stable $CsPbBr_3$ quantum dots by silica-coating and ligand modification for white light-emitting diodes and visible light communication. *Chem. Eng. J.* **2021**, *419*, 129551. [CrossRef]
81. Yang, L.; Fu, B.; Li, X.; Chen, H.; Li, L. Poly(vinylidene fluoride)-passivated CsPbBr3 perovskite quantum dots with near-unity photoluminescence quantum yield and superior stability. *J. Mater. Chem. C* **2021**, *9*, 1983–1991. [CrossRef]
82. Song, J.; Fang, T.; Li, J.; Xu, L.; Zhang, F.; Han, B.; Shan, Q.; Zeng, H. Organic–Inorganic Hybrid Passivation Enables Perovskite QLEDs with an EQE of 16.48%. *Adv. Mater.* **2018**, *30*, 1805409. [CrossRef] [PubMed]
83. Yang, F.; Chen, H.; Zhang, R.; Liu, X.; Zhang, W.; Zhang, J.; Gao, F.; Wang, L. Efficient and Spectrally Stable Blue Perovskite Light-Emitting Diodes Based on Potassium Passivated Nanocrystals. *Adv. Funct. Mater.* **2020**, *30*, 1908760. [CrossRef]
84. Jang, J.; Kim, Y.-H.; Park, S.; Yoo, D.; Cho, H.; Jang, J.; Jeong, H.B.; Lee, H.; Yuk, J.M.; Park, C.B.; et al. Extremely Stable Luminescent Crosslinked Perovskite Nanoparticles under Harsh Environments over 1.5 Years. *Adv. Mater.* **2021**, *33*, 2005255. [CrossRef] [PubMed]
85. Gerion, D.; Pinaud, F.; Williams, S.C.; Parak, W.J.; Zanchet, D.; Weiss, S.; Alivisatos, A.P. Synthesis and Properties of Biocompatible Water-Soluble Silica-Coated CdSe/ZnS Semiconductor Quantum Dots. *J. Phys. Chem. B* **2001**, *105*, 8861–8871. [CrossRef]
86. Xie, H.; Chen, E.; Ye, Y.; Xu, S.; Guo, T. Highly Stabilized Gradient Alloy Quantum Dots and Silica Hybrid Nanospheres by Core Double Shells for Photoluminescence Devices. *J. Phys. Chem. Lett.* **2020**, *11*, 1428–1434. [CrossRef]
87. Xie, H.; Chen, E.; Ye, Y.; Xu, S.; Guo, T. Interfacial optimization of quantum dot and silica hybrid nanocomposite for simultaneous enhancement of fluorescence retention and stability. *Appl. Phys. Lett.* **2020**, *117*, 171101. [CrossRef]
88. Stöber, W.; Fink, A.; Bohn, E. Controlled growth of monodisperse silica spheres in the micron size range. *J. Colloid Interface Sci.* **1968**, *26*, 62–69. [CrossRef]
89. Ziegler, J.; Xu, S.; Kucur, E.; Meister, F.; Batentschuk, M.; Gindele, F.; Nann, T. Silica-Coated InP/ZnS nanocrystals as converter material in white LEDs. *Adv. Mater.* **2008**, *20*, 4068–4073. [CrossRef]
90. Hu, Z.; Liu, Z.; Bian, Y.; Li, S.; Tang, X.; Du, J.; Zang, Z.; Zhou, M.; Hu, W.; Tian, Y.; et al. Enhanced Two-Photon-Pumped emission from in situ synthesized nonblinking $CsPbBr_3/SiO_2$ nanocrystals with excellent stability. *Adv. Opt. Mater.* **2018**, *6*, 1700997. [CrossRef]
91. Chen, W.; Hao, J.; Hu, W.; Zang, Z.; Tang, X.; Fang, L.; Niu, T.; Zhou, M. Enhanced stability and tunable photoluminescence in perovskite $CsPbX_3$/ZnS quantum dot heterostructure. *Small* **2017**, *13*, 1604085. [CrossRef]
92. Li, Z.; Hofman, E.; Li, J.; Davis, A.H.; Tung, C.; Wu, L.; Zheng, W. Photoelectrochemically active and environmentally stable $CsPbBr_3/TiO_2$ Core/Shell nanocrystals. *Adv. Funct. Mater.* **2018**, *28*, 1704288. [CrossRef]
93. Cai, J.; Lin, J.; Chen, Y.; Xu, S.; Ye, Y.; Chen, E.; Guo, T. Stacked Encapsulation Structure for Discretely Distributed Quantum Dot Array. *IEEE Photonics J.* **2020**, *12*, 1–10. [CrossRef]
94. Chen, Y.; Cai, J.; Lin, J.; Hu, X.; Wang, C.; Chen, E.; Sun, J.; Yan, Q.; Guo, T. Quantum-dot array with a random rough interface encapsulated by atomic layer deposition. *Opt. Lett.* **2022**, *47*, 166–169. [CrossRef] [PubMed]
95. Xiang, Q.; Zhou, B.; Cao, K.; Wen, Y.; Li, Y.; Wang, Z.; Jiang, C.; Shan, B.; Chen, R. Bottom up Stabilization of $CsPbBr_3$ Quantum Dots-Silica Sphere with Selective Surface Passivation via Atomic Layer Deposition. *Chem. Mater.* **2018**, *30*, 8486–8494. [CrossRef]
96. Wang, H.; Lin, S.; Tang, A.; Singh, B.P.; Tong, H.; Chen, C.; Lee, Y.; Tsai, T.; Liu, R. Mesoporous silica particles integrated with All-Inorganic $CsPbBr_3$ perovskite Quantum-Dot nanocomposites (MP-PQDs) with high stability and wide color gamut used for backlight display. *Angew. Chem. Int. Ed.* **2016**, *55*, 7924–7929. [CrossRef]
97. Dirin, D.N.; Protesescu, L.; Trummer, D.; Kochetygov, I.V.; Yakunin, S.; Krumeich, F.; Stadie, N.P.; Kovalenko, M.V. Harnessing Defect-Tolerance at the nanoscale: Highly luminescent lead halide perovskite nanocrystals in mesoporous silica matrixes. *Nano Lett.* **2016**, *16*, 5866–5874. [CrossRef]
98. Malgras, V.; Henzie, J.; Takei, T.; Yamauchi, Y. Stable blue luminescent CsPbBr3 perovskite nanocrystals confined in mesoporous thin films. *Angew. Chem.* **2018**, *130*, 9019–9023. [CrossRef]
99. Yang, G.; Fan, Q.; Chen, B.; Zhou, Q.; Zhong, H. Reprecipitation synthesis of luminescent $CH_3NH_3PbBr_3/NaNO_3$ nanocomposites with enhanced stability. *J. Mater. Chem. C* **2016**, *4*, 11387–11391. [CrossRef]
100. Dirin, D.N.; Benin, B.M.; Yakunin, S.; Krumeich, F.; Raino, G.; Frison, R.; Kovalenko, M.V. Microcarrier-Assisted inorganic shelling of lead halide perovskite nanocrystals. *ACS Nano* **2019**, *13*, 11642–11652. [CrossRef]
101. Wei, Y.; Xiao, H.; Xie, Z.; Liang, S.; Liang, S.; Cai, X.; Huang, S.; Al Kheraif, A.A.; Jang, H.S.; Cheng, Z.; et al. Highly luminescent lead halide perovskite quantum dots in hierarchical CaF_2 matrices with enhanced stability as phosphors for white Light-Emitting diodes. *Adv. Opt. Mater.* **2018**, *6*, 1701343. [CrossRef]
102. Shen, J.; Wang, Y.; Zhu, Y.; Gong, Y.; Li, C. A polymer-coated template-confinement $CsPbBr_3$ perovskite quantum dot composite. *Nanoscale* **2021**, *13*, 6586–6591. [CrossRef]
103. Cai, Y.; Li, Y.; Wang, L.; Xie, R.-J. A Facile Synthesis of Water-Resistant $CsPbBr_3$ Perovskite Quantum Dots Loaded Poly(methyl methacrylate) Composite Microspheres Based on In Situ Polymerization. *Adv. Opt. Mater.* **2019**, *7*, 1901075. [CrossRef]
104. Lin, J.; Wang, Z.; Zhang, R.; Huang, F.; Chen, D. Research progresses in preparation and applications of CsPbX 3 (X= Cl, Br, I) perovskite quantum dots-embedded glass. *Chin. J. Lumin.* **2021**, *42*, 1331–1344. [CrossRef]
105. Hou, J.; Chen, P.; Shukla, A.; Krajnc, A.; Wang, T.; Li, X.; Doasa, R.; Tizei, L.H.G.; Chan, B.; Johnstone, D.N.; et al. Liquid-phase sintering of lead halide perovskites and metal-organic framework glasses. *Science* **2021**, *374*, 621–625. [CrossRef]
106. Chen, E.; Xie, H.; Huang, J.; Miu, H.; Shao, G.; Li, Y.; Guo, T.; Xu, S.; Ye, Y. Flexible/curved backlight module with quantum-dots microstructure array for liquid crystal displays. *Opt. Express* **2018**, *26*, 3466–3482. [CrossRef]

107. Lin, S.; Tan, G.; Yu, J.; Chen, E.; Weng, Y.; Zhou, X.; Xu, S.; Ye, Y.; Yan, Q.F.; Guo, T. Multi-primary-color quantum-dot down-converting films for display applications. *Opt. Express* **2019**, *27*, 28480–28493. [CrossRef]
108. Chen, E.; Lin, J.; Yang, T.; Chen, Y.; Zhang, X.; Ye, Y.; Sun, J.; Yan, Q.; Guo, T. Asymmetric Quantum-Dot Pixelation for Color-Converted White Balance. *ACS Photonics* **2021**, *8*, 2158–2165. [CrossRef]
109. Huang, B.; Chen, E.; Sun, L.; Guo, T. Quantum-dot color conversion film patterned by screen printing and overprinting process for display backlights. *Opt. Laser Technol.* **2022**, *145*, 107486. [CrossRef]
110. Huang, B.L.; Guo, T.L.; Xu, S.; Ye, Y.; Chen, E.G.; Lin, Z.X. Color Converting Film With Quantum-Dots for the Liquid Crystal Displays Based on Inkjet Printing. *IEEE Photonics J.* **2019**, *11*, 1–9. [CrossRef]
111. Yin, Y.; Meng, H. Progress of quantum dots and perovskite as color conversion materials for full-color display. *Chin. J. Lumin.* **2021**, *42*, 419–447. [CrossRef]
112. Zhang, X.; Chen, A.; Yang, T.; Cai, J.; Ye, Y.; Chen, E.; Xu, S.; Ye, Y.; Sun, J.; Yan, Q.; et al. Tripling Light Conversion Efficiency of μLED Displays by Light Recycling Black Matrix. *IEEE Photonics J.* **2022**, *14*, 1–7. [CrossRef]
113. Wang, C.; Cai, J.; Ye, Y.; Hu, X.; Zhong, L.; Xie, H.; Chen, E.; Ye, Y.; Xu, S.; Sun, J.; et al. Full-visible-spectrum perovskite quantum dots by anion exchange resin assisted synthesis. *Nanophotonics* **2022**, *11*, 1355–1366. [CrossRef]
114. Nick, R.J.; Breen, C.A.; Denton, C.M.; Sadasivan, S.; Linton, J.R.; QD Vision Inc. Method of making components including quantum dots, methods, and products. U.S. Patent Application 20150049491A1, 19 February 2015.
115. Ji, H.; Ye, D.; Xu, H.; Chen, E.; Ge, Z. Multi-layer co-extruded quantum-dot diffuser plate for ultra-large TV backlights. *Opt. Mater. Express* **2022**, *12*, 1648–1656. [CrossRef]
116. Chen, E.; Guo, J.; Jiang, Z.; Shen, Q.; Ye, Y.; Xu, S.; Sun, J.; Yan, Q.; Guo, T. Edge/direct-lit hybrid mini-LED backlight with U-grooved light guiding plates for local dimming. *Opt. Express* **2021**, *29*, 12179–12194. [CrossRef]
117. Wang, J.; Ye, Y.; Xu, S.; Kang, D.; Lin, Z.; Chen, E.; Guo, T.; Yang, L. Fabrication and properties of quantum-dots backlight light guide plate. *Chin. J. Lumin.* **2017**, *38*, 91–96. [CrossRef]
118. Chen, E.; Lin, S.; Jiang, Z.; Guo, Q.; Xu, S.; Ye, Y.; Yan, Q.F.; Guo, T. Analytic design of light extraction array for light guide plate based on extended sources. *Opt. Express* **2019**, *27*, 34907–34920. [CrossRef]
119. Yang, H.S.; Noh, S.H.; Suh, E.H.; Jung, J.; Oh, J.G.; Lee, K.H.; Jang, J. Enhanced stabilities and production yields of MAPbBr3 quantum dots and their applications as stretchable and Self-Healable color filters. *ACS Appl. Mater. Interfaces* **2021**, *13*, 4374–4384. [CrossRef]
120. Cai, J.; Wang, C.; Hu, X.; Ye, Y.; Zhong, L.; Chen, E.; Ye, Y.; Xu, S.; Sun, J.; Yan, Q.; et al. Water-driven photoluminescence reversibility in $CsPbBr_3$/PDMS-PUa composite. *Nano Res.* **2022**. [CrossRef]
121. Yin, Y.; Ali, M.U.; Liu, M.; Miao, J.; Peng, W.; Li, D.; Chen, S.; Lee, C.; Meng, H. Vacuum-drying processed micrometer-thick stable $CsPbBr_3$ perovskite films with efficient blue-to-green photoconversion. *Small* **2019**, *15*, 1901954. [CrossRef]
122. Hu, Z.; Yin, Y.; Ali, M.U.; Peng, W.; Zhang, S.; Li, D.; Zou, T.; Li, Y.; Jiao, S.; Chen, S.J.; et al. Inkjet printed uniform quantum dots as color conversion layers for full-color OLED displays. *Nanoscale* **2020**, *12*, 2103–2110. [CrossRef]
123. Yan, F.; Xing, J.; Xing, G.; Quan, L.; Tan, S.T.; Zhao, J.; Su, R.; Zhang, L.; Chen, S.; Zhao, Y.; et al. Highly efficient visible colloidal Lead-Halide perovskite nanocrystal Light-Emitting diodes. *Nano Lett.* **2018**, *18*, 3157–3164. [CrossRef] [PubMed]
124. Han, D.; Imran, M.; Zhang, M.; Chang, S.; Wu, X.; Zhang, X.; Tang, J.; Wang, M.; Ali, S.; Li, X.; et al. Efficient Light-Emitting diodes based on in situ fabricated FAPbBr3 nanocrystals: The enhancing role of the Ligand-Assisted reprecipitation process. *ACS Nano* **2018**, *12*, 8808–8816. [CrossRef]
125. Huang, H.; Zhao, F.; Liu, L.; Zhang, F.; Wu, X.; Shi, L.; Zou, B.; Pei, Q.; Zhong, H. Emulsion synthesis of Size-Tunable $CH_3NH_3PbBr_3$ quantum dots: An alternative route toward efficient Light-Emitting diodes. *ACS Appl. Mater. Interfaces* **2015**, *7*, 28128–28133. [CrossRef] [PubMed]
126. Xing, J.; Yan, F.; Zhao, Y.; Chen, S.; Yu, H.; Zhang, Q.; Zeng, R.; Demir, H.V.; Sun, X.; Huan, A.; et al. High-Efficiency Light-Emitting diodes of organometal halide perovskite amorphous nanoparticles. *ACS Nano* **2016**, *10*, 6623–6630. [CrossRef] [PubMed]
127. Cho, H.; Jeong, S.; Park, M.; Kim, Y.; Wolf, C.; Lee, C.; Heo, J.H.; Sadhanala, A.; Myoung, N.; Yoo, S.; et al. Overcoming the electroluminescence efficiency limitations of perovskite light-emitting diodes. *Science* **2015**, *350*, 1222–1225. [CrossRef]
128. Zhang, X.; Liu, H.; Wang, W.; Zhang, J.; Xu, B.; Karen, K.L.; Zheng, Y.; Liu, S.; Chen, S.; Wang, K.; et al. Hybrid perovskite Light-Emitting diodes based on perovskite nanocrystals with Organic-Inorganic mixed cations. *Adv. Mater.* **2017**, *29*, 1606405. [CrossRef]
129. Zhang, X.; Han, D.; Wang, C.; Muhammad, I.; Zhang, F.; Shmshad, A.; Xue, X.; Ji, W.; Chang, S.; Zhong, H. Highly efficient light emitting diodes based on in situ fabricated $FAPbI_3$ nanocrystals: Solvent effects of on-chip crystallization. *Adv. Opt. Mater.* **2019**, *7*, 1900774. [CrossRef]
130. Pan, G.; Bai, X.; Xu, W.; Chen, X.; Zhai, Y.; Zhu, J.; Shao, H.; Ding, N.; Xu, L.; Dong, B.; et al. Bright blue light emission of Ni^{2+} ion-doped $CsPbCl_xBr_{3-x}$ perovskite quantum dots enabling efficient light-emitting devices. *ACS Appl. Mater. Interfaces* **2020**, *12*, 14195–14202. [CrossRef]
131. Kim, Y.; Lee, G.; Kim, Y.; Wolf, C.; Yun, H.J.; Kwon, W.; Park, C.G.; Lee, T.W. High efficiency perovskite light-emitting diodes of ligand-engineered colloidal formamidinium lead bromide nanoparticles. *Nano Energy* **2017**, *38*, 51–58. [CrossRef]
132. Li, X.; Yang, L.; Yang, Q.; Wang, S.; Ding, J.; Wang, L. Heterogeneous post-passivation of inorganic cesium lead halide perovskite quantum dots for efficient electroluminescent devices. *J. Mater. Chem. C* **2021**, *9*, 3978–3986. [CrossRef]

133. Khan, Q.; Subramanian, A.; Yu, G.; Maaz, K.; Li, D.; Sagar, R.U.R.; Chen, K.; Lei, W.; Shabbir, B.; Zhang, Y. Structure optimization of perovskite quantum dot light-emitting diodes. *Nanoscale* **2019**, *11*, 5021–5029. [CrossRef] [PubMed]
134. Dong, Y.; Wang, Y.; Yuan, F.; Johnston, A.; Liu, Y.; Ma, D.; Choi, M.; Chen, B.; Chekini, M.; Baek, S.; et al. Bipolar-shell resurfacing for blue LEDs based on strongly confined perovskite quantum dots. *Nat. Nanotechnol.* **2020**, *15*, 668–674. [CrossRef] [PubMed]
135. Tsai, H.; Shrestha, S.; Vilá, R.A.; Huang, W.; Liu, C.; Hou, C.; Huang, H.; Wen, X.; Li, M.; Wiederrecht, G.; et al. Bright and stable light-emitting diodes made with perovskite nanocrystals stabilized in metal–organic frameworks. *Nat. Photonics* **2021**, *15*, 843–849. [CrossRef]

MDPI AG
Grosspeteranlage 5
4052 Basel
Switzerland
Tel.: +41 61 683 77 34

Nanomaterials Editorial Office
E-mail: nanomaterials@mdpi.com
www.mdpi.com/journal/nanomaterials

Disclaimer/Publisher's Note: The title and front matter of this reprint are at the discretion of the Guest Editors. The publisher is not responsible for their content or any associated concerns. The statements, opinions and data contained in all individual articles are solely those of the individual Editors and contributors and not of MDPI. MDPI disclaims responsibility for any injury to people or property resulting from any ideas, methods, instructions or products referred to in the content.

www.ingramcontent.com/pod-product-compliance
Lightning Source LLC
Chambersburg PA
CBHW041953260926
51850CB00019B/2686
* 9 7 8 3 7 2 5 8 4 6 0 9 2 *